Electronic Instruments and Measurements

Patrick Crozier
Quincy Vocational-Technical School and
Quincy Junior College

DELMAR PUBLISHERS INC.®

Cover photo courtesy of Hewlett-Packard Company

For information, address Delmar Publishers Inc.,
2 Computer Drive West, Box 15-015
Albany, New York 12212

Printed in the United States of America
Published simultaneously in Canada
By Nelson Canada
A Division of The Thomson Corporation

10 9 8 7 6 5 4 3 2 1

ISBN 0-8273-3874-0

To Susan, Diane, Patty, and Sally

Contents

8 OSCILLOSCOPES   151

13 ANALYZERS   282

14 AUTOMATIC TEST EQUIPMENT   301

Preface

Electronic technicians are employed to do, not just to know. Much of the doing involves analyzing circuit performance through the use of measurement instruments. This task can be simple and require little accuracy, as in determining whether a circuit is alive. It can also be complex and require many accurate measurements and some mathematical analysis, as in determining the probable value of a resistor randomly selected from an assortment of like-coded resistors. Thus, measurement is more than connecting two leads to a circuit and reading a scale. It is also the selection of instruments, the establishment of procedures, and the preparation of documentation.

This text was written to help students understand the importance of measurement in electronics and to acquire the knowledge, skills, and attitudes necessary to become competent in using measurement instruments and procedures. For measurement, a knowledge of electronics theory is necessary but is not an end in itself. Theory is presented if it is needed for an understanding of the operation, capabilities, and limitations of an instrument. Similarly, mathematical procedures are presented when they help explain concepts or are part of the analysis process.

The text begins with definitions of essential terms, such as measurement, accuracy, reliability, and error. Electrical parameters, units, and standards are described. Relevant mathematical procedures are discussed, including the calculation of mean, median, mode, and standard deviation. Error types and analysis are discussed. The need for and methods of data documentation are explained and demonstrated.

The theory of operation, the characteristics, and the applications of measurement instruments are described next. Traditional instruments such as analog meters, bridges, oscilloscopes, and potentiometers are included, along with digital meters and the more recent network analyzers and computer-interfaced, automatic test equipment. Instrument calibration is the final topic.

Student exercises require definition, calculation, comparison, and analysis. Some laboratory activities are suggested for concept reinforcement. These activities are specific in task but not in equipment, since all labs may not have the same equipment available.

The author has intentionally avoided the extensive theory typically included in some measurement texts for two reasons. First, it incorrectly implies that the student will need to know how to design these circuits. In reality, the student will most often just use them. Second, the need for design competence is often one step too far for an otherwise employable technician. In reality, more time will be spent working outside the cabinet using instruments in their intended applications rather than inside the cabinet modifying its design. With this in mind, the author recommends that this text be reinforced with manufacturers' catalogs and instruction manuals so that the student becomes experienced in the comparison and selection of instruments and in the habit of following the proper operating instructions. Of course, the most important reinforcement needed is extensive and varied lab measurement work.

High-quality measurements are not always needed, but technicians must be able to make them when required. This text was written to help students develop that ability. The ultimate test is their future performance in measuring the parameters of electronic devices and circuits in a safe, efficient, and accurate manner and in effectively communicating the results to others.

Many people provided assistance in the preparation of this book. Thanks are extended to those who reviewed and commented upon preliminary drafts of the manuscript for this book: Alexander W. Avtgis, Wentworth Institute of Technology, MA; James R. Shelton, Hawkeye Institute of Technology, IA; Frederick M. Sawyer, Rhode Island Junior College, RI; Walter Luke, Jr., Broome County Community College, NY; Antonio A. Migale, Moorpark College, CA; George Mathis, Spartanburg Technical College, SC.

Special thanks and gratitude are extended to Edward P. Kearney of Waterbury State Technical College, CT, for his extremely thorough and thoughtful review of the final draft and to the production staff of Breton Publishers—especially Carol Beal, whose skillful editing immeasurably improved the presentation.

1

Measurement Process

OBJECTIVES

This chapter is an introduction to the field of electronics measurement. The purpose of measurement, some common terms, and some factors that technicians must consider are discussed.

Name three dimensions of safety in measurement.
Describe the terms *measurement, accuracy, precision,* and *reliability.*
Name factors to consider before beginning any measurements.
Name factors to consider while conducting measurements.
Describe tasks to be performed following measurements.

INTRODUCTION

Historical Overview

People have long been connecting meters and other instruments to electric circuits in order to make a determination about operating conditions. Thirty years ago, instruments were large vacuum tube–operated units that required time to warm up and stabilize. Equipment was more cumbersome and less accurate than equipment of today. Measurement procedures were awkward and time-consuming. Specifications and standards were not clearly defined. Data analysis procedures required considerable time since electronic computers were not so readily available. The field of measurement has improved dramatically since then.

Today, most technicians spend part of their day using electronic instruments to measure voltage, frequency, and other circuit characteristics. Equipment is more accurate, durable, convenient, and complicated than its counterpart of a decade ago. Because applications for electronic equipment are becoming more critical with each passing day, so, too, are measurement requirements. Thus, circuit and component specifications are more clearly defined, and standards of today demand higher-quality products.

Today's trends in measurement—increased accuracy, greater precision, instrument miniaturization, and computerization—will continue at an ever increasing rate. Greater production efficiency from the demand for reduced manufacturing costs will lead to more automated test and measurement systems. Integrated circuits will lead to even smaller instruments with increased accuracy, ease of use, and flexibility of operation. Computer connection will provide for computer control, monitoring, and data analysis. New measurement applications will evolve.

The role of the electronic technician continues to include the selection and use of electronic instruments for the safe and accurate determination of circuit characteristics. This role requires an understanding of instrument capabilities and limitations, a knowledge of circuit theory, and a recognition of and aim for quality work. Yet, the list of requirements does not end here, for as circuits, components, equipment, and standards change, so must the technician. Thus, the technician needs to make a commitment to remain aware of the state of the art in this rapidly changing segment of the electronics industry.

Important Terms

The language of electronics includes some terms used in the area of measurement, the most important of which is *safety*. More will be said about safety in a moment. Other terms are *measurement, test, parameter, error, accuracy,* and *resolution.* Few of these terms have only one definition. Instead, they are defined by individuals or groups who use them to suit specific situations. For this reason, the definitions used in this text combine the common uses. In the future, when you are working in industry, you will find many situations where a company has its own specifications and procedures.

Measurement can be defined as a process in which an instrument is used to quantify—that is, to give a numerical value to—a parameter. A *parameter* is a circuit dimension or element, such as voltage, power, or resistance. A *test* is a process of observation or examination. One major difference exists between test and measurement. In measurement, a specific number or value is determined. Tests need not lead to numbers.

If a radio is not working, you can use a meter to determine the voltage level at the wall outlet. Another approach might be to connect a lamp to the outlet to see whether the circuit is alive. The difference between these two methods is that the first involves quantification and the second does not. *To quantify* means to determine a numerical value or size, and that is what measurement is all about. An *instrument* is a device used to quantify an unknown parameter. A thermometer is a typical instrument; it is used to quantify temperature. Typical electronic instruments are meters, oscilloscopes, and bridges.

Two of the most important characteristics of an instrument are accuracy and precision. *Accuracy* describes how close to the true value an instrument will indicate. *Precision* is the degree of resolution or the number of significant figures an instrument will give. Accuracy and precision are independent of one another but equally essential.

For every parameter, there is an absolute value at the time of measurement. The closer an instrument indication is to this true value, the more accurate the instrument is. A clock that differs 2 minutes (abbreviated min) from a time standard is less accurate than one that differs by 2 seconds (s). The latter clock has less error than the former clock, since *error* can be defined as the difference between indicated and true values.

A clock that can indicate seconds is more precise than a clock that indicates only hours and minutes. It could, however, be wrong. The clock without a second hand may have the correct time but be hard to read without a second hand; hence it may lead to reader error. The former clock is precise but not accurate, while the latter clock is accurate but not precise. Once again, both accuracy and precision are essential for high-quality measurements.

Reliability is defined as the degree of consistency with which an instrument performs. If a meter indicates 2.4 volts (abbreviated V) now, will it indicate 2.4 volts later if the voltage does not change? It should. Variations should only be caused by the parameter. Thus, a person who arrives 10 minutes late for work every day is reliable because you can bank on the consistency. Unfortunately, this person is not desirable because he or she is not accurate about arrival time.

Decisions in Measuring

In general, the measurement process involves the following steps:

1. Decide to measure.
2. Select a procedure.
3. Conduct the measurement.
4. Analyze the data.
5. Report the results.

Sometime in the course of a day, you may need to conduct a measurement. It could be a routine measurement with specific directions, perhaps a measurement that you have made often and effectively. We are not concerned with that situation here. Our concern is the measurement you have not made before, a measurement of a parameter or a particular circuit you may be unfamiliar with.

You have a few decisions to make before you connect an instrument in order to make a measurement. These decisions are based on questions that must be answered. What parameter is being measured? What are the characteristics of this parameter? What quality is required? While the decision to measure the parameter is most likely made by someone else, the procedures you will follow are probably determined by you. The decisions you make will influence the quality of your measurements, and *quality is the primary consideration* in measurement. Other factors you should consider when selecting a measurement procedure are discussed in a later section.

Specific Value of a Parameter

You have probably measured some parameter in the past and obtained a variety of different results. Variation in results can occur for several reasons. One possibility is that the equipment was defective. Another possibility is that a parameter might have changed between measurements. Or the instrument may be doing what it is designed to do yet be incapable of what you expect of it.

Quite often, a parameter varies between measurements. If it does, the variation cannot be ignored. Your final data must include a description of this variation. Details of how you might describe the variation are presented in Chapter 3. Once again, though, each parameter in an electronic system does have specific value at any given instant, and that value can be measured. Like a truth, the value does exist. Your task is to determine it and define it.

SELECTING A PROCEDURE

The major factors involved in the selection of a procedure are as follows:

> Characteristics of the parameter,
> Number of measurements,
> Quality required,
> Instruments and options,
> Safety.

A discussion of these factors follows.

Characteristics of the Parameter

Obviously, before you make a measurement, you must know what parameter you are to measure and where in the circuit it is located. For good measurements, however, this knowledge is not enough. You also need to know some of the characteristics of the parameter.

If you are going to measure an alternating current (AC) voltage, you must know its approximate frequency in order to select an instrument capable of responding to that frequency. All AC voltmeters do not operate over the same frequency range. You also need to know about the waveshape, harmonics, or direct current (DC) offset, for these factors can affect results. It may seem ironic that you are asked to describe some characteristics of a parameter *before* you begin to measure it. However, that is what you must do. Thus, the knowledge of parameter characteristics that you gained in other electronics courses will be important for your measurement work in this course.

When measuring frequency with an electronic counter, as shown in Figure 1–1, you could damage the counter by using too large a signal. In this case, one characteristic of a signal is creating a problem while you are measuring another characteristic. Discussions of the capabilities and limitations of different measuring instruments constitute a major part of later chapters.

Quality Required

Quality requirements—that is, the time and costs that are budgeted for a project—are major factors in the determination of measurement procedures. A cost is involved in any measurement process.

───○ **FIGURE 1–1**
Frequency Measurement with an Electronic Counter (Photo
courtesy of Hewlett-Packard Company)

Both time and equipment can be expensive. How much time should
you spend to arrive at a result? What quality of equipment is required? So, in selecting a measurement procedure, you must make
a cost-value decision. It is not reasonable to use the best equipment
available and take as much time as you want if you are just trying
to decide if a circuit is alive or if the voltage is somewhere around
117 volts.

    Very often, important decisions involving a great deal of money
will be based on the results of your measurements. For example,
you may be assigned to evaluate a prototype of a new product for
your company. The results of your evaluation will form the basis of
a management decision to manufacture 1000 units just like this one.
What if you are wrong? Your information would cause an expensive
loss in time and materials for your company. This problem is generally avoided if you have prior knowledge of the importance and

value of your measurements and results. This knowledge helps you determine the time and equipment limitations or costs.

Instruments and Options

Another decision you need to make involves the selection of the equipment you will use. If you are in a small lab, you may have little choice. Generally, though, you will have a number of different products to choose from. You can measure voltage with a meter or an oscilloscope. You can measure frequency with an oscilloscope or a counter. Which instrument should you choose? How do you decide? Once again, you need to know the parameter, the quality required, the equipment available, and the capabilities of that equipment.

While an oscilloscope allows you to see the shape of a voltage, it does not have the accuracy of a meter. It can, however, measure smaller voltages than a meter can. Also, oscilloscopes are not as accurate as counters for frequency measurement; yet, they can indicate some circuit characteristics that would upset a counter. In other words, knowledge of the characteristics of both the instrument and the parameter is important in your selection of measurement equipment.

Establishing a Procedure

Once you know what you are to measure and what you will measure it with, you must decide *how* to measure. This decision involves more than connecting an instrument and turning on the power. For example, are any disconnections necessary? If so, plan to make them with the power turned off. Then, connect the instrument, turn the power on if necessary, and proceed.

You also must decide how many measurements to make. Is one enough? If the parameter is likely to change, you may want to plan on more than one measurement. How many will be enough? Your knowledge of the parameter will help you decide how many measurements you need.

Finally, what will you record, and where will you record it? You should plan to have a notebook available; do not expect to remember your results. You may also want to summarize or otherwise process your data. This topic is considered in Chapter 3. Documentation of what you did and how you did it is also important. Measurement records and reports will be discussed in Chapter 4.

Once you know just what you are going to do and have established a procedure to do it, you can proceed with your measurements.

CONDUCTING THE MEASUREMENT

When your procedures are established, you are almost ready to begin. But before you energize any equipment, remember that electricity is dangerous and equipment is expensive. You attitude toward safety and the care of equipment is important. Basically, you should follow these steps:

1. Read the manual.
2. Work safely.
3. Collect data.
4. Be patient.
5. Question results.

Work Habits

The quality of your results will be affected by your work habits. Valuable instruments need proper care. Limitations must be recognized and accounted for. Procedures should be safe for you, the equipment, and the people around you. Patience is needed while you are gathering data. Neatness and accuracy are essential in the preparation of reports and other documentation. Furthermore, you must always be conscious of safety and aim for quality.

One more characteristic is needed if you are to be good at measurement: You need to question. Have I selected the best instrument? Is this the correct location to find the parameter? Have I taken enough measurements in order to have representative data? And a final question often ignored: Do these results seem believable? When you can answer yes to these questions, the success of your efforts will be more probable.

Safety First

There are three dimensions involved in safe work practices: you, the test equipment being used, and the items being measured. The primary danger to you is electric shock. Shock can generally be avoided if you know the voltages present and give them the attention they command.

The test equipment you are working with can be damaged by overloading or overdriving. Once again, know the limitations of what you are working with and observe them. Read the manual that accompanies the equipment.

Damage to what you are working on must also be avoided. Equipment does not always come with a manual listing the ways that damage can occur during testing. For example, a transmitter can be damaged if operated without an antenna. Integrated circuits can often be damaged if tested out of a circuit. Furthermore, some measurement instruments, while not causing damage, can change the operation of a circuit being evaluated, causing erroneous results. Your best protection is to know what you are working with, what you are working on, and what you are doing. Then, just be careful.

Sampling

A problem mentioned earlier was the variation or change in a parameter during measurement. How do you know that a parameter will not change? You don't. How do you know which value to choose when a parameter keeps changing? The answer is, You cannot pick just one value. You must take enough measurements to ensure that you have a representative sample. Then, summarize the data by some procedure such as averaging. (This topic is discussed in Chapter 3.)

Once again, you must know how important the answer is and how much time and effort you can afford to spend in its pursuit. No single number can be offered as the absolute number of measurements needed. However, measuring a line voltage over a 24-hour period, for instance, should provide reasonable data for a description, since most factors affecting its value should occur during that period. The characteristics of a box of like-coded resistors can be reasonably described by taking a random sample. Of course, the most accurate method would be to measure all the resistors, but that procedure is usually not reasonable.

TASKS AFTER THE MEASUREMENT

The process is not completed when the measurements are taken. Equipment must be returned to its proper place, and your results must be communicated to all interested parties. Data must be analyzed and summarized, and results must be reported.

Analysis of Data

If your total effort consisted of one measurement, there is little more to do. Just report the number. However, this simple situation rarely arises. Usually, you will have a considerable amount of data that needs to be summarized. The data might come from a series of measurements over a period of time or over a number of components. While you may report this information on a chart or a graph, that method is often inadequate. A mathematical summary is often better.

One mathematical method of data analysis is averaging; however, it is limited as a description. For example, line voltages of 115, 115, and 115 produce the same average as voltages of 100, 115, and 130. No one would ever report that these two situations are the same. Thus, other methods are necessary to clearly describe what is happening. Some common analysis methods will be described in Chapter 3.

Reporting Results

Generally, results are reported in writing so that details can be communicated clearly and accurately. Another reason for written reports is that many projects involve many measurements over a period of days or even weeks. A report provides a historical record of what was done, why, how, and by whom.

Most of your reports will describe your intentions, your procedures, your results, and your conclusions. The equipment you used is identified by name, brand, model, and serial number. Your procedures are outlined. The data you gathered are presented in tables, charts, or graphs, and samples of all calculations are presented. Questions are answered. Finally, your conclusions are stated at the end. While all other parts of a report are facts, conclusions are opinions. Basically, your conclusions are a result of a comparison between what was intended or expected and what was observed. It is also a summary statement that can be supported by the facts in your report. Further aspects of reports are discussed in Chapter 4.

Decisions

Another process begins once your results have been reported. Someone will make decisions on the basis of your report. Is the component as specified? Is the circuit working as expected? Can we build 1000 more units just like this one? These questions will lead to decisions that, as mentioned before, can be costly. Often, the person responsible for making these decisions will be your boss, and he or she will be held accountable if the decision proves to be wrong. In turn, you will be held accountable for the results you report.

So, we return to an earlier point: You must know what degree of quality is required in your measurements. If you need to know whether a circuit is alive, the question can be resolved with a simple test. However, if you need to know the voltage variation within 0.1% over a 24-hour period, a more difficult and costly process is involved. Thus, you should try to find out, in advance, the type of decisions that will be made so that you can adjust your measurement procedures accordingly.

SUMMARY

The measurement process has seen both change and stability over the past few decades. Test equipment has become more accurate, more precise, more durable, and less cumbersome as a result of microelectronics. Yet, the terms used in the procedure are still the same. Measurement is a process in which an instrument is used to quantify a parameter. A test is a process of observation or examination. Accuracy describes how close to the true value an instrument indicates. Precision is the degree of resolution of the measurement, while reliability is the degree of consistency in results.

The decision to take a measurement is based on the realization that a true or absolute value does exist for a parameter. Your task is to quantify it or place a numerical value on it. The decision to measure, often made by others, leads to a number of other decisions you must make, such as what equipment you should use and how many measurements you should make. These decisions are influenced by the degree of quality required, a fact you need to know before you begin.

Equipment selection is an important step. You must understand the specifications of an instrument before you select it. Thus, you should read the manual before you begin. Once you begin, you should always consider your own safety and the safety of those around you, of the instrument you are working with, and of the product you are working on. Also, in your procedure, you will need to take enough measurements to ensure that you can adequately describe the parameter under investigation. Finally, as you proceed, you need to keep asking yourself, Is this the best I can do to quantify this variable?

When the measurement is completed, you must report your results. Usually, you will supply a written report describing your intentions, procedures, data, results, and conclusions. Also, you may need to process the data in order to provide a summarized description of the parameter.

Decisions to make design changes, to manufacture more units like the one tested, or to conduct more tests are usually made from the information in technicians' reports. The costs associated with these decisions can be quite high, and they are directly affected by the quality of your performance. That is why we said earlier that quality is the primary consideration in measurement.

BIBLIOGRAPHY

The growing need for instrumentation technicians, Technical Education News, March–April, 1981.

Where are we going in test and measurement? How will we get there? Test & Measurement World, February–March, 1982.

Test instrument safety, who cares? Electronic Products, March 28, 1983.

Instrumentation trends, smaller and smarter is better, Evaluation Engineering, March, 1984.

Testing for military applications, a tough customer with megabucks, Evaluation Engineering, July, 1984.

Guide to Electronic Measurements and Laboratory Practice, 2nd ed., Stanley Wolf, Prentice-Hall, 1983.

QUESTIONS

1. Describe three dimensions of safety you should consider in your measurement procedure.
2. Define *measurement*.
3. Define *instrument*.
4. Define *accuracy*.
5. Define *precision*.
6. Define *reliability*.
7. Describe some trends in the field of measurement.
8. Give several reasons why people measure parameters.
9. What is meant by the statement that a value does exist?

10. Why is it necessary to know the characteristics of a parameter before you measure it?

11. Why is it necessary to know the degree of quality required for a measurement?

12. What is meant by the phrase *cost versus quality?*

13. Why is it necessary to know the specifications of an instrument before you use it?

14. What is meant by *sampling?* Why is sampling done?

15. What is reliability, and why is it a concern?

16. Describe four work habits that affect the quality of your work.

17. What methods are commonly used to report results? Why?

18. Why is averaging an inadequate method of describing variation?

2

Parameters, Units, and Standards

OBJECTIVES

This chapter reviews basic electrical parameters, their units, and their derivations. Standards for these units are introduced and discussed. Common prefixes and their conversions are also presented.

> Describe the International System of Units.
> Define the terms *fundamental unit, derived unit, standard,* and *decade.*
> Name the seven fundamental units and their corresponding parameters.
> Convert values from one prefix designation to another.
> Identify and describe four levels of standards.

SYSTEMS OF UNITS

Parameters, units, and standards are used to name and describe quantities in electric and electronic circuits. A *unit* is the name given to a parameter's dimension. For example, the meter is a unit used for length. A *standard* is a universally accepted description or definition of a unit's magnitude. While thousands of different circuits can be assembled from the multitude of components in existence, their characteristics are described with a universal terminology. A system of units provides the foundation for this language.

Need for Systems

Electronics is both a national and an international industry. Without a universal system of units, communication between vendors, manufacturers, and consumers, often separated by great distances, could lead to misunderstandings and nonconformity of components. For example, would a 1-ohm resistor manufactured in one state be considered 1 ohm in another? The concern is not that the component would change value but, rather, that source and destination measurement processes would not be the same. That is, what one person calls 1 ohm, another person may not.

We can expand this concern to other quantities or parameters such as voltage, current, frequency, and time. All persons using the term *volt* should give it the same meaning. A quantity called 1 volt at one location should be equivalent to the quantity called 1 volt at all other locations. Thus, if successful trade is to exist, universally accepted definitions are required. People cannot play a game successfully unless all persons in the game use the same rules.

Development of Systems

The need for uniformity in the description of electrical quantities was considered over a hundred years ago by the British Association for the Advancement of Science. This organization established a committee to identify a practical unit for resistance and to develop a way of defining its magnitude. The result was a universal description and physical representation of the magnitude that all interested parties could use as a standard. Another type of standard was also established, the *physical standard,* which is a physical representation of the defined standard.

The initial physical standard for resistance was a coil of platinum-silver alloy wire sealed in a paraffin- (wax) filled container. This physical standard matched the definition for 1 ohm of resistance. Around 1900, the definition and physical standard were changed to the resistance of a column of mercury of specific length, diameter, and temperature. Now, the resistance definition is in terms of length, mass, and time, and the physical representation is a bank of sealed wire-wound resistors.

Meanwhile, units and standards were defined for other electrical parameters. Current was defined as the amount of silver deposited by current flowing through a silver nitrate solution in a given amount of time. The unit was called the international ampere. Electromotive force was defined in terms of the international ampere, the international ohm (mercury column) described earlier, and Ohm's law. These standards were the status of the system of units around the year 1900.

International System of Units

For years, different systems of units were used throughout the world. The foot-pound-second system (FPS system) was used in the United States, while the *meter-kilogram-second system* (MKS system) was used in Europe. Then, increased international trade of manufactured goods demonstrated the need for uniformity of standards and units. Also, technological developments provided both the need for higher-quality standards and the capability of fabricating them.

    Le Système International d'Unités, or *International System of Units* (SI system), was developed in France in 1960 and accepted by 36 countries. The MKS system is part of it. The International Committee on Weights and Measures (CGPM) meets every six years to consider and recommend improvements in the system. With SI, each and every quantity has a unit name, symbol, and defined magnitude. Definitions of magnitude use reproducible physical phenomena. This system is the one in use today.

UNITS

Fundamental Units

The foundation of SI is built on seven basic or *fundamental units* and two supplementary units. All other units are called *derived units* since they are expressed in terms of two or more of these nine units. Table 2–1 presents the names, symbols, and definitions of the SI fundamental and supplementary units.

Derived Units

As just noted, derived (or compound) units are described in terms of two or more fundamental units. Like descriptive standards, physical standards for derived units can be established in terms of two or more fundamental physical standards. A number of common derived units are listed in Table 2–2.

──○ **TABLE 2–1**

Names and Symbols of SI Fundamental and Supplementary Units

Fundamental Units

Quantity	Name of Unit	Symbol
Length	Meter	m
Mass	Kilogram	kg
Time	Second	s
Electric current	Ampere	A
Thermodynamic temperature	Kelvin	K
Luminous intensity	Candela	cd
Amount of substance	Mole	mol

Supplementary Units

Quantity	Name of Unit	Symbol
Plane angle	Radian	rad
Solid angle	Steradian	sr

──○ **TABLE 2–2**

Names and Symbols of SI Derived Units

Quantity	Name of Unit	Symbol	Derivation
Area	Square meter	m^2	
Volume	Cubic meter	m^3	
Frequency	Hertz	Hz	s^{-1}
Mass density (density)	Kilogram per cubic meter	kg/m^3	
Speed, velocity	Meter per second	m/s	
Angular velocity	Radian per second	rad/s	
Acceleration	Meter per second squared	m/s^2	
Angular acceleration	Radian per second squared	rad/s^2	
Force	Newton	N	$kg\text{-}m/s^2$
Pressure (mechanical stress)	Pascal	Pa	N/m^2

(continued)

───○ **TABLE 2–2**
(continued)

Quantity	Name of Unit	Symbol	Derivation
Kinematic viscosity	Square meter per second	m^2/s	
Dynamic viscosity	Newton-second per square meter	Ns/m^2	
Work, energy, quantity of heat	Joule	J	Nm
Power	Watt	W	J/s
Quantity of electricity	Coulomb	C	As
Potential difference, electromotive force	Volt	V	W/A
Electric field strength	Volt per meter	V/m	
Electric resistance	Ohm	Ω	V/A
Capacitance	Farad	F	As/V
Magnetic flux	Weber	Wb	Vs
Inductance	Henry	H	VsA
Magnetic flux density	Tesla	T	Wb/m^2
Magnetic field strength	Ampere per meter	A/m	
Magnetomotive force	Ampere	A	
Luminous flux	Lumen	lm	cd-sr
Luminance	Candela per square meter	cd/m^2	
Illuminance	Lux	lx	lm/m^2
Wave number	1 per meter	m^{-1}	
Entropy	Joule per kelvin	J/K	
Specific heat capacity	Joule per kilogram-kelvin	J/(kg-K)	
Thermal conductivity	Watt per meter-kelvin	W/(m-K)	
Radiant intensity	Watt per steradian	W/sr	
Activity (of a radioactive source)	1 per second	s^{-1}	

Electrical Parameters

The most common electrical parameter you will measure is voltage. The volt (V) is the unit used to describe a quantity of electromotive force or potential difference. As you can see in Table 2–2, the volt is derived from the ampere and the watt, and the watt is derived from the joule and the second. Furthermore, the joule is derived from the newton and the meter, and the newton is derived from the meter, the kilogram, and the second. In other words, the unit volt can be traced back to the fundamental units of meter, kilogram, second, and ampere. The definitions of some other common electrical parameters are also given in Table 2–2.

Prefixes and Conversions

Prefixes are words used to represent a unit multiplier; they simplify the expression and manipulation of data. For example, the SI prefix *mega-* is used to represent the word *million* or the quantity 1,000,000. Its symbol is M. Thus, a 1,500,000-ohm resistor is a 1.5 megohm resistor, written as 1.5 MΩ (Ω is the Greek capital letter omega). Likewise, the prefix *micro-* is used to represent the term *one-millionth* or the quantity 0.000001. Its symbol is the Greek letter μ (mu). Thus, a current of 0.000003 ampere is 3 microamperes and is written a 3 μA. Common SI prefixes are presented in Table 2–3.

The use of prefixes makes calculations easier, especially when we are using scientific notation (powers of ten).

○ **Example of Scientific Notation**

In scientific notation, 1,500,000 ohms can be written as 1.5×10^6, and 0.000003 ampere can be written as 3×10^{-6}. The voltage in a circuit with these parameters is calculated, without prefixes, in this way:

$$V = IR = 3 \times 10^{-6}\,A \times 1.5 \times 10^6\,\Omega = 4.5\,V$$

All electronics formulas are based on units without prefixes— that is, on *absolute units*. Converting from absolute units to units with prefixes can be accomplished by using the factors listed in Table 2–3. As shown in the table, prefix values are in steps of three decimal places. So, we just move the decimal point of the quantity to be redescribed 3, 6, 9, or 12 places to a location giving the smallest number equal to or larger than one. Then, we write the power of

───○ **TABLE 2–3**
SI Prefixes

Factor by Which Unit Is Multiplied	Prefix	Symbol
10^{12}	Tera	T
10^9	Giga	G
10^6	Mega	M
10^3	Kilo	k
10^2	Hecto	h
10	Deka	da
10^{-1}	Deci	d
10^{-2}	Centi	c
10^{-3}	Milli	m
10^{-6}	Micro	μ
10^{-9}	Nano	n
10^{-12}	Pico	p
10^{-15}	Femto	f
10^{-18}	Atto	a

ten, with the exponent representing the number of places moved. Finally, we eliminate the power of ten expression and replace it with the appropriate prefix.

○ Example of Converting from Absolute Units to Units with Prefixes

Convert 2200 ohms and 0.015 ampere to units with prefixes.

$$2200\,\Omega = 2.2 \times 10^3\,\Omega = 2.2\,k\Omega$$

$$0.015\,A = 15 \times 10^{-3}\,A = 15\,mA$$

Converting in the opposite direction is done in a similar way. Just write the expression with the prefix, change the prefix to its appropriate power of ten, and multiply.

○ Example of Converting from Units with Prefixes to Absolute Units

Convert 2200 picofarads and 0.25 milliampere to absolute units.

$$2200\,pF = 2200 \times 10^{-12}\,F = 0.000000002200\,F$$

$$0.25\,mA = 0.25 \times 10^{-3}\,A = 0.00025\,A$$

One safe way to convert from one prefix to another is through absolute units. While you may easily remember that a particular prefix represents three decimal places, you may get confused about direction. As a result, your answer could be six decimal places off. In converting with absolute units, we use what is called the unity conversion factor. This factor is simply a fractional expression in which the numerator and the denominator are equal; thus, the fraction equals one (or unity). Terms used in the fraction are selected so that they cancel existing units and introduce desired units. The following example illustrates the technique.

○ Example of Converting through Absolute Units

Convert 120 microamperes to milliamperes.

$$120\,\mu A \times \frac{1\,A}{1,000,000\,\mu A} \times \frac{1000\,mA}{1\,A} = 0.12\,mA$$

Convert 0.5 megawatt to kilowatts.

$$0.5\,MW \times \frac{1,000,000\,W}{1\,MW} \times \frac{1\,kW}{1000\,W} = 500\,kW$$

PHYSICAL STANDARDS

Physical standards are components, devices, or systems that maintain or produce accurate and precise magnitudes of quantities. They are used for calibrating or verifying the accuracy of measuring instruments, and they exist for all SI quantities. The cost of physical standards can range from under $100 to more than $100,000 and is directly related to accuracy and precision. Since, in many applications, the highest quality of standards is not required or is too costly, different levels of standards exist.

Levels of Standards

There are basically four levels of standards:

> International,
> Primary,
> Secondary,
> Working.

International standards are those maintained at the International Bureau of Weights and Measures in France. They are of the highest quality possible and are absolutely accurate, by definition. Most international standards consist of systems that produce the quantity, as described by the defined standard. The kilogram standard is the only one represented by a physical object. International standards are used for reference with national primary standards and are inaccessible to the public.

    Primary standards are kept in selected locations in the world, such as the National Bureau of Standards (NBS) in Washington, D.C., and the National Physical Laboratory in England. These standards are regularly compared and corrected toward their mean in order to maintain consistent primary standards. *Secondary standards* are also kept at these locations and are used for comparison of primary and working standards. *Working standards* are used in industrial laboratories in the day-to-day process of calibrating instruments.

Primary Standards

The primary standard for time at the NBS is a cesium beam atomic clock, which can define the second with an accuracy of 1×10^{-13}. The use of this clock is consistent with the definition of the second, which is based on the duration of 9,192,631,770 periods of oscillation of a cesium-133 atom. Since frequency and time are inversely related, the time standard is the basis for the frequency standard.

    The NBS *radio stations WWV and WWVB* broadcast pulsed tones of accurately controlled frequencies and durations. The audio frequencies, radio frequencies, and time increments available are described in the Appendix. These standards are available in any laboratory through the use of a communications receiver.

    The resistance standard at NBS provides an accuracy and precision greater than 0.06 parts per million (ppm). The standard volt and standard ohm are used to provide the standard ampere. Capacitance is established to better than 0.02 ppm, and inductance is given to better than 0.06 ppm.

Secondary Standards

The secondary, or reference, standards are the highest level of standards a manufacturer would have, and they are compared periodically with standards at NBS. Secondary standards do not generally have the performance capabilities of the primary standards at NBS. Manufacturers rarely need such capabilities.

Reference standards are used in the calibration of working standards. Reference standards are maintained in a temperature-, humidity-, and dust-controlled environment and are used only by calibration technicians. In a typical application, reference standards are removed once a month and used to calibrate the working standards. Once every year or so, the reference standard is taken to NBS for calibration. Although secondary standards see very little service, they are very important in industry.

Working Standards

A technician's major contact is with working standards. Figure 2–1 shows a DC current and voltage standard with a voltage accuracy of 0.003% of setting, a resolution of 1 part per million, and a 1-hour (h) stability of 0.00075%. The output voltage spans from 0.1 microvolt to 10 volts over three ranges, and a controlled environment is not required. Current is available from 0.01 microampere to 100 milliamperes in three ranges. Current accuracy is 0.005%, and the 1-hour stability is 0.001%.

The standard capacitor in Figure 2–2 is made of a silver-mica and foil pile mounted in a cast aluminum case. Capacitors of this

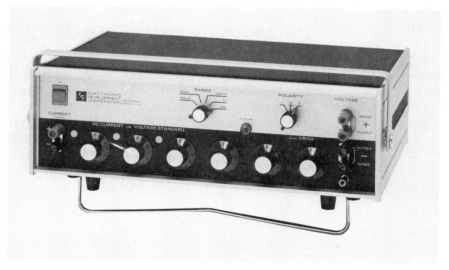

———○ **FIGURE 2–1**
DC Current and Voltage Standard (Photo courtesy of Electronic Development Corporation)

———O **FIGURE 2–2**
Standard Capacitor (Photo courtesy of GenRad, Inc.)

type come in values from 0.001 to 1 microfarad and are available with an accuracy of 0.02% and a stability of 0.01% per year.

The standard inductor in Figure 2–3 consists of a toroid wound on a ceramic core and is mounted in ground cork in an aluminum case. Inductors of this type come in values from 100 microhenrys to 10 henrys. Adjustment accuracy is 0.1%, and stability is 0.01% per year.

Figure 2–4 shows a standard resistor intended for use in a laboratory or production area for calibrating devices such as resistance bridges. Low-resistance standard resistors are made from low–temperature coefficient sheet metal, with holes that reduce inductance. Higher-resistance standards use various types of wire. All are sealed in oil-filled containers, which give them long-term stability. Individual values are available in ranges from 0.01 ohm to 1 megohm. Accuracy ranges from 0.01% to 0.1%, and stability is 10 parts per million per year.

Decades

The resistors, inductors, and capacitors just described are all single-value standards. Multiple-value standards are also available. These standards are usually provided by *decades,* which are devices that

─────○ **FIGURE 2–3**
Standard Inductor (Photo courtesy of GenRad, Inc.)

─────○ **FIGURE 2–4**
Standard Resistor (Photo courtesy of GenRad, Inc.)

give values in steps of 1, 10, 100, and so on. For example, a five-dial decade resistor can indicate resistances from 0 to 999,990 ohms in 10-ohm steps. This device has an accuracy of 0.01% and five-place precision. A five-place decade inductor can provide inductances up to 99.999 henrys in 0.001-henry steps. A precision, six-place decade capacitor can produce values from 0 to 9.99999 microfarads at an accuracy of 0.05%. Because of their wide selection of accurate and precise values, decades give technicians a convenient way to calibrate instruments at a number of different operating points.

SUMMARY

The International System of Units (SI) defines quantities in terms of a unit name, a symbol, and a magnitude. For example, the quantity for electric current has a unit name of ampere and the symbol A. The magnitude of 1 ampere is defined in terms of the force between parallel current-carrying conductors under specifically described conditions.

Units are divided into three categories: fundamental, supplementary, and derived. The seven fundamental units represent the quantities of length, time, mass, electric current, temperature, luminous intensity, and amount of substance. The two supplementary units describe the quantities of plane angle and solid angle. All remaining units are called derived units because they are derived from the first nine. For example, the unit hertz for frequency is derived from time.

Except for the ampere, all electrical parameters are derived. The watt is derived from the joule and the second; the joule is derived from the newton and the meter. The volt is derived from the ampere and the watt. The newton is derived from the meter, the kilogram, and the second. While you can measure parameters such as power and voltage directly, most characteristics you will investigate will be derived.

Physical standards serve as representations of the defined magnitude of quantities. The kilogram, representing mass, is the only quantity represented by an actual object. All other fundamental units are represented by systems that produce a measurable phenomenon. For example, the United States primary standard for time is a cesium beam atomic clock. It has a relative accuracy exceeding 1 microsecond.

The highest level of standard is the international standard. These standards are kept in France. The next level consists of primary standards, which are kept at locations such as the NBS in Washington, D.C. Secondary, or reference, standards are the next level, and they are generally used as the standards in industrial applications. The working standard is the lowest level and is used in the day-to-day calibration of other instruments. It is compared with a secondary standard according to a calibration schedule. Likewise, the secondary standard is regularly compared with a primary standard for calibration.

Many working standards for resistance, capacitance, and inductance are of fixed value, but adjustable standards are also available. Adjustable standards, called decade resistors, decade inductors, and decade capacitors, provide technicians with useful means of instrument calibration.

BIBLIOGRAPHY

International System of Units (SI), E. A. Mechtly, NASA, 1973.
Bushbaum's Complete Handbook of Practical Electronic Reference Data, 2nd ed., Prentice-Hall, 1978.
Fundamentals of Physical Measurement, Ernest Zebrowski, Jr., Duxbury Press, 1979.
Electronic Engineer's Handbook, 2nd ed., McGraw-Hill Publishing Company, 1982.

QUESTIONS

1. Define *fundamental units*.
2. Define *derived units*.
3. Define *standard*.
4. Define *decade*.
5. Explain why uniform standards are essential.
6. Describe the International System of Units.
7. Identify four levels of standards. Describe their characteristics and applications.
8. Name the seven fundamental units and their corresponding parameters.
9. Which electrical parameter has a fundamental unit? How is it defined?

10. Upon what fundamental units is the quantity of resistance derived? Describe specifically how it is derived.

11. Describe the standards available from WWV and WWVB.

12. Compare the applications of secondary and working standards.

13. Compare working standards and decades. What is the primary advantage of decades?

14. Name four common SI prefixes. Indicate their magnitudes in absolute numbers and with scientific notation.

15. What is the CGPM, and what does it do? Since standards have already been established, why is it necessary for this organization to meet on a regular basis?

16. The FPS system continues to be used in the United States, even though the MKS system has been adopted essentially worldwide. Why do you think the FPS system is still used so much in the United States?

17. Time standards are the most accurate and readily available standards. What makes them so accurate?

18. What type of unit is the volt? What is it based on?

19. What type of unit is the ohm? What is it based on?

20. What type of unit is the watt? What is it based on?

PROBLEMS

1. Express the following values as numbers between 1 and 10 by using SI prefixes: 0.00024 ampere, 750,000 ohms, 0.00000002 farad, 35,000,000 ohms.

2. Express the following values as numbers between 1 and 10 by using SI prefixes: 2,700,000 ohms, 18,600 ohms, 0.0335 ampere, 0.0000047 ampere.

3. Express the following values as numbers between 1 and 10 by using SI prefixes: 27,000 ohms, 0.015 volt, 0.0105 ampere, 0.0002 second.

4. Express the following values in absolute units: 1.5 kilovolts, 150 millivolts, 4.72 microvolts, 0.47 megohm.

5. Express the following values in absolute units: 3.4 megohms, 0.25 milliampere, 5 microvolts, 0.350 kilovolt.

6. Express the following values in absolute units: 12 nanoseconds, 0.025 picofarad, 12.5 kilohms, 0.05 volt.

7. Convert the following values as described, using SI prefixes: 0.35 milliampere to microamperes, 750 volts to kilovolts, 2200 picofarads to microfarads.

8. Convert the following values as described using SI prefixes: 330 picofarads to microfarads, 1.5 milliamperes to microamperes, 470 kilohms to megohms.

9. Convert the following values as described, using SI prefixes: 440 microamperes to milliamperes, 0.052 megohm to kilohms, 1500 milliwatts to watts.

3

Methods of Data Analysis

OBJECTIVES

The measurement of component and circuit parameters often produces a considerable amount of data, and you will have to process these data in order to produce an accurate and descriptive summary of parameters. This chapter deals with data analysis by reviewing familiar mathematical procedures and introducing a few new ones.

Name some differences between nominal, ordinal, interval, and ratio scales.

Determine the number of significant figures in a number.

Name and describe five types of error.

Calculate the limiting error of a group of numbers.

Determine the mean, median, mode, and midrange of a group of numbers.

Determine the standard deviation and the normal curve distribution of a group of numbers.

LEVELS OF MEASUREMENT

Numbers are used both to identify and to describe components and circuits. They can be manipulated through a range of processes, from finding the sum or difference to finding the variance and standard deviation. However, all processes cannot be applied to all numbers. The process appropriate for the data depends on the level of measurement of the numbers. Four levels of measurement are described here, along with mathematical procedures allowed for each.

Nominal Scales

The first or lowest level of measurement is the *nominal scale,* and it includes those numbers that identify or classify items. Examples are your Social Security number and telephone number. No meaningful calculations, not even addition or subtraction, can be performed with these numbers. The only procedure you can perform with them is count how many there are.

Ordinal Scales

Ordinal scales are at the second level of measurement, and they indicate rank or order. The list of the top ten records is an ordinal scale. This list indicates that hit #1 is above #2 and that they both are above #3. Since there is no indication of the magnitude of difference between these records, the only mathematical process allowable is what has already been done, rank ordering.

Interval Scales

The third level of measurement is the *interval scale,* and it indicates the order of the numbers as well as the magnitude of difference between individual points. Furthermore, interval scales have a zero reference, a point at which the numerical value of zero (0) is positioned. The Fahrenheit (°F) and Celsius (°C) temperature scales are two examples. A 1-degree increase from 35 to 36 degrees Celsius is equal in magnitude to a 1-degree rise from 42 to 43 degrees. Also, the Celsius scale has a zero reference point.

However, in interval scales, one characteristic is still missing if we are to perform calculations. The zero reference point must indicate the complete absence of the physical characteristic being measured. In the Celsius scale, for instance, 0 degrees Celsius is an arbitrary point; it does not indicate a complete absence of heat.

Ratio Scales

The fourth and final scale is the *ratio scale.* In it, numbers indicate order and the distance between each other. Furthermore, each point can be directly related to a zero reference point that indicates complete absence of the physical characteristic. The kelvin (K) and Rankine (°R) temperature scales are ratio scales. In these scales, zero represents a complete absence of heat.

Fortunately, in electronics, parameters such as voltage, current, resistance, and power are measured on ratio scales. Data derived from ratio scales can be subjected to all calculations, without limitation.

Significant Figures

The number of *significant figures* indicates the degree of resolution or certainty to which a magnitude is known. Significant figures are the basis for precision, which is interdependent with accuracy in clearly defining true values.

All digits except 0 are significant all of the time. The digit 0 is significant when it is between two nonzero digits, is to the right of the decimal part of a number, or has an overbar.

○ Examples of Significant Figures

Some examples of significant figures are given in the following table.

One Significant Figure	Four Significant Figures
7	9257
200	703.1
0.003	52.03
6×10^3	$800.\overline{6} \times 10^6$
$\overline{3}900$	$400\overline{0}$

The concept of significant figures leads to an important rule in calculations. *The number of significant figures in a result cannot exceed the number of significant figures in the factor with the fewest significant figures.*

○ Example of Finding the Significant Figures in a Product

Determine the number of significant figures in the answer to 35.5 times 0.25.

$$35.5 \times 0.25 = 8.875$$

Since the second factor has only two significant figures, the answer must have only two significant figures. So, the answer is as follows:

$$35.5 \times 0.25 = 8.8$$

Notice in the example that the extra digits were not used to round off the answer. Since they had no significance, they had to be dropped without consideration. A similar situation occurs with a sum of numbers.

○ Example of Finding the Significant Figures in a Sum

The following addition shows how a sum is expressed with the correct number of significant figures.

$$125.3 + 147.25 + 113. + 123.78 + 130.5 = 638.$$

The third number has only three significant figures. The first unknown digit to the right of the decimal point could be 9 or it could be 1. Since it is unknown, its impact on the column is also unknown. For that reason, the column must be ignored and the answer expressed in three significant figures.

○ Examples of Significant Figures for Other Calculations

The results in the following calculations are expressed with the correct number of significant figures.

$$0.0020 \times 3.28 = 0.0065$$

$$\frac{57.9}{24} = 2.4$$

$$250 + 187 + 17.7 = 454$$

$$1425 - 250.3 = 1175$$

ERROR

Error is the difference between an actual or true value and its indicated value. The role of a technician is to identify all errors and eliminate them. Errors that cannot be eliminated must be compensated for through measurement or analysis procedures.

Types of Error

The error type determines the resolution method. Errors can be categorized in various ways, such as by cause, as in instrument errors, observer errors, or parameter errors. They can also be categorized by behavior, such as systematic errors or random errors.

Systematic errors are those that recur consistently, with the same magnitude. *Instrument errors,* which are errors caused by the instrument used, are usually systematic errors. For example, an uncalibrated instrument will continually give the same wrong answer. Worn parts and temperature change are additional causes of systematic error. You can reduce the effects of these errors by using a different instrument, stabilizing the temperature, or introducing the error factor into your calculations.

Another systematic error is *observer error,* an error caused by the person conducting the measurement. Observer error could occur, for instance, when you measure AC voltage above the upper frequency limits of a volt-ohm meter (VOM). The instrument is not at fault because it is doing what its specifications say it can. You are at fault because of your selection. Other causes of observer error are misreading and selecting the wrong range. All observer errors are avoidable.

Random error is error that is both unpredictable and inconsistent in its presence. Random error is usually caused by *parameter error,* which is an error occurring because of some abnormal phenomenon in the parameter. Transients in a voltage that you are measuring are an example. For instance, you may measure the power line voltage just as a sudden load drops the level to 113 volts, although its is almost always 118 volts. Your conclusion of 113 volts would be incorrect because that value was a random rather than a typical magnitude. The impact of random errors can be reduced with multiple measurements and statistical analysis.

Limiting Error

Manufacturer specifications may indicate that a 100-ohm resistor has a tolerance of $\pm 10\%$; this figure gives maximum limits of error. *Limiting error* is the maximum error expected from individual or combined sources. It is the result of combining the worst case of all individual errors. For the resistor just mentioned, the manufacturer guarantees that the *individual* resistance will be within 10% of 100 ohms—that is, somewhere between 90 and 110 ohms.

Now, let us consider what happens when resistors are combined, as shown in the next example.

○ Example of Limiting Error When Resistors Are Combined

Here, we will consider two 1000-ohm resistors, one with 5% tolerance and one with 10% tolerance. The following table gives the initial calculations (R is the symbol for a resistor or a resistance).

Quantity	R_1	R_2
Ideal values	1000 Ω	1000 Ω
Maximum error	1000 Ω × 0.05 = 50 Ω	1000 Ω × 0.10 = 100 Ω
Minimum value	1000 Ω − 50 Ω = 950 Ω	1000 Ω − 100 Ω = 900 Ω
Maximum value	1000 Ω + 50 Ω = 1050 Ω	1000 Ω + 100 Ω = 1100 Ω

Now, let us consider the compound error when these resistors are connected in series. Remember that when we consider compound error, we *always consider the worst case.* The worst case here would be if both resistors were at their low limits or both were at their high limits. The calculations are as follows, where R_T denotes total resistance:

$$\text{minimum value of } R_T = \text{minimum } R_1 + \text{minimum } R_2$$
$$= 950\ \Omega + 900\ \Omega = 1850\ \Omega$$

$$\text{maximum value of } R_T = \text{maximum } R_1 + \text{maximum } R_2$$
$$= 1050\ \Omega + 1100\ \Omega = 2150\ \Omega$$

In both cases, the error is 150 ohms, since the error is equal to the difference between the indicated and actual values.

$$\text{error for minimum} = 2000\ \Omega - 1850\ \Omega = 150\ \Omega$$
$$\text{error for maximum} = 2150\ \Omega - 2000\ \Omega = 150\ \Omega$$

Error is usually stated as a percentage of the indicated value and can be found in this way:

$$\% \text{ error} = \frac{\text{error}}{\text{indicated value}} \times 100\% = \frac{150}{2000} = 7.5\%$$

The important rule to note here is as follows: *Rule 1—Errors in sums are averaged according to weight.*

There is another rule about errors: *Rule 2—Errors in products and quotients are added algebraically.* For instance, the impact of the percentage of variation of a resistor depends on the basic value of that resistor and whether it is connected in series or in parallel with other resistors. In series connection, a 1-megohm, 10% resistor would have a far greater effect than a 1-kilohm, 20% resistor. Yet, the 1-kilohm resistor would have the greater impact if the connection were parallel.

○ Example of Power Calculation and Rule 2

Let's consider a power calculation for two measurements, 50.0 volts and 0.200 ampere, both guaranteed to be within ±3%. From these measurements, we are expected to calculate circuit power, circuit resistance, and limiting error. Our calculations would begin with the determination of error in each parameter, as follows:

$$3\% \text{ of } 50.0 \text{ V} = 0.03 \times 50.0 \text{ V} = 1.50 \text{ V}$$

$$3\% \text{ of } 0.200 \text{ A} = 0.03 \times 0.200 \text{ A} = 0.006 \text{ A}$$

Now, we can determine the power. *The worst case in a product occurs when both values are high or when both are low.* We will use both high values here. The indicated power is as follows:

$$\text{indicated power} = 50 \text{ V} \times 0.200 \text{ A} = 10 \text{ W}$$

However, if we assume the maximum possible error, the power could be much higher, as shown next.

$$\text{maximum voltage} = 50.0 \text{ V} + 1.5 \text{ V} = 51.5 \text{ V}$$

$$\text{maximum current} = 0.200 \text{ A} + 0.006 \text{ A} = 0.206 \text{ A}$$

$$\text{maximum power limit} = (\text{voltage} + \text{error}) \times (\text{current} + \text{error})$$
$$= (50.0 \text{ V} + 1.5 \text{ V}) \times (0.200 \text{ A} + 0.006 \text{ A})$$
$$= 51.5 \text{ V} \times 0.206 \text{ A} = 10.6 \text{ W}$$

$$\text{error} = \text{actual value} - \text{indicated value} = 10.6 \text{ W} - 10.0 \text{ W}$$
$$= 0.6 \text{ W}$$

$$\% \text{ error} = \frac{\text{error}}{\text{indicated value}} \times 100\% = \frac{0.6}{10.0} \text{ W} = 6\%$$

As stated in Rule 2, the two 3% errors combine to produce a 6% error.

This impact is also true for quotients, as we can see when resistance is calculated with the values from the previous example. There is one change in our procedure. *Worst-case values in a quotient occur when one factor is at its upper limit and the other is at its lower limit.*

○ Example of Resistance Calculation and Rule 2

We will calculate the indicated resistance, the maximum limit, and the error.

$$\text{indicated resistance} = \frac{50\text{ V}}{0.200\text{ A}} = 250\ \Omega$$

$$\text{maximum limit: } = \frac{50\text{ V} + 1.5\text{ V}}{0.200\text{ A} - 0.006\text{ A}} = \frac{51.5\text{ V}}{0.194\text{ A}} = 265\ \Omega$$

$$\%\text{ error} = \frac{\text{error}}{\text{indicated value}} \times 100\% = \frac{15\ \Omega}{250\ \Omega} = 6\%$$

As the calculations show, the two 3% errors combine to produce a 6% error.

Errors will not always be as large as the errors in the examples. A 3% meter is not necessarily 3% off, for instance. If you know the actual error, you can introduce it into your calculations and provide a measured value with the qualifying limiting error. However, *if you do not know the actual error in a measurement, always assume it to be the worst case.*

MEASURES OF CENTRAL TENDENCY

You may have taken many measurements of a variable and want to summarize your results with one number. When you reach this point, averaging usually comes to mind. You believe that the most representative number is the one in the middle. But where is the center of all of your measurements? It could be the average, although that is not the only summary number available for describing the *central tendency* of a group of measurements. Four methods will be explained here.

Mean

The *mean,* or average, is the most common measure of central tendency. Its calculation can be demonstrated with a simple example.

○ Example of Calculating the Mean

Consider a small company of nine employees. We would like summarize their annual incomes. We begin by finding the mean $\overline{X}$, which is equal to the sum of all values, Σx, divided by the number of values n. (Σ is the Greek letter sigma and denotes "sum.")

$$\text{mean} = \frac{\sum x}{n}$$

The salaries appear in Table 3–1. Thus, the mean is calculated as follows:

$$\text{mean} = \frac{\sum x}{n} = \frac{\$225,000}{9} = \$25,000$$

We can state that the average annual salary of the employees in this company is $25,000. Since only one of the nine people earns that exact amount, the mean can be misleading.

───○ **TABLE 3–1**
Salaries of Nine Employees Used to Calculate the Mean

Person	Salary
A	$15,000
B	25,000
C	10,000
D	60,000
E	26,000
F	18,000
G	15,000
H	21,000
I	35,000
Total	$225,000

Median

The *median* is the central number in a list given in rank order. We find the median in the next example.

○ Example of Finding the Median

To find the median, we rearrange the salaries listed in Table 3–1 from smallest to largest. The rearranged salaries are given in Table 3–2. The middle income (the fifth of nine values) is $21,000. But once again, only one person makes that amount of money. This number is also misleading.

Mode

The *mode* is the number that appears most often. Look again at Table 3–2. It indicates that if you were to ask the employees how much they earn, more would say $15,000 than any other amount. So, the mode of this group is $15,000.

Midrange

Another measure of central tendency is the *midrange*, the center between the highest and the lowest number.

────○ **TABLE 3–2**
Salaries in Rank Order Used to Calculate the Median

Person	Salary	
C	$10,000	
G	15,000	
A	15,000	
F	18,000	
H	21,000	Median
B	25,000	
E	26,000	
I	35,000	
D	60,000	

○ **Example of Calculating the Midrange**

As Table 3–2 shows, the salaries cover a very wide range. The range is from $10,000 to $68,000. The midrange is determined in the following manner:

$$\text{midrange} = \frac{\text{smallest} + \text{largest}}{2} = \frac{\$10,000 + \$60,000}{2} = \$35,000$$

Once again, we have a salary that represents the income of only one individual.

Table 3–3 shows a summary of our results, presenting each measure of central tendency for the salaries listed in Table 3–1. So, which number do we use? Our choice depends on what we want to say. The mode value of a box of resistors tells us the size we are most likely to get if we randomly choose one from a box. The midrange of a line voltage indicates the center between the high-voltage and the low-voltage point. Of all of the measures of central tendency, though, the mean is the most useful for describing what the parameter is or what has occurred. However, we must occasionally use other methods, such as measures of variation.

MEASURES OF VARIATION

The preceding section of this chapter demonstrated how difficult it can be to describe the magnitude of a parameter that changes. The

──○ **TABLE 3–3**
Summary of Results for Salaries, Showing Mode, Median, Mean, and Midrange

Salary	Measure
$10,000	
15,000	
15,000	Mode
18,000	
21,000	Median
25,000	Mean
26,000	
35,000	Midrange
60,000	

mean seems to be the most representative value. The fault with the mean is that it ignores the degree of variation that almost always exists. For example, line voltages of 105, 120, and 135 volts have the same mean value as voltages of 118, 120, and 122 volts. Yet, the first circuit is far from being identical to the second. *Variability,* or the differences of individual measurements, needs to be included in any description of this voltage.

Measurement for high-volume production applications requires that samples of a parameter or product be taken in order to predict all values of it and to account for variability. While this procedure is typically carried out by computerized, automatic test equipment, you should be familiar with the process.

Standard Deviation

The *standard deviation* is the root-mean-square (rms) value of variation and is used with the mean to describe parameters. It is calculated by using the following steps:

1. Calculate the mean $\overline{X}$ of the individual values x.
2. Determine the individual deviations d—that is, the distance of each value from the mean.
3. Square the deviations (d^2).
4. Find the sum of the deviations squared (Σd^2).
5. Divide the sum of the deviations squared by the number of values n minus one.
6. Find the square root.

The final answer is called the standard deviation σ (Greek lowercase letter sigma) and has the same units as the parameter.

○ Example of Calculating the Standard Deviation

The standard deviation of a group of 15 resistors is determined in the following manner. First, as shown in Table 3–4, we find the individual deviations and the deviations squared of the sample resistors. Then, we determine the mean as follows:

$$\overline{X} = \frac{\Sigma x}{n} = \frac{15,030\,\Omega}{15} = 1002\,\Omega$$

──○ **TABLE 3–4**
Individual Values, Deviations, and Deviations Squared of
Sample Resistors Used to Calculate Standard Deviation

Individual Value (x)	Deviation (d)	Deviation Squared (d^2)
985	−17	289
998	−4	16
1,014	12	144
1,004	2	4
925	−77	5,929
950	−52	2,704
1,078	76	5,776
1,026	24	576
1,001	−1	1
1,090	88	7,744
996	−6	36
932	−70	4,900
1,051	49	2,401
1,008	6	36
972	−30	900
$\Sigma x = 15,030$		$\Sigma d^2 = 31,456$

Finally, we determine the standard deviation:

$$\sigma = \sqrt{\frac{\Sigma d^2}{n - 1}} = \sqrt{\frac{31,456}{14}} = 47.4 \, \Omega$$

The mean resistance value is 1002 ohms, and the standard deviation
is 47.4 ohms. In order to realize the significance of this number, we
must take one more step. This step is described in the next subsection.

Probability and the Normal Curve

All of our calculations so far have been about what is, or conditions
as they now exist. We have described the characteristics of all items
measured by using what is called *descriptive statistics*. The subject
we are about to consider is called *inferential statistics*. With inferential statistics, we attempt to infer, or predict, the description of
values we have not measured.

Let's imagine that the 15 resistors of the previous example were a sample taken from a box of 500. While 15 is a rather small sample, it is adequate for a description of the mathematical process; a sample of 30 would be more reasonable, in practice. What conclusions can we draw from the calculations of the example? While we did not measure all resistors, we can attempt to predict what the characteristics of the others *probably* are—hence the term *probability*. Probability is defined as the mathematical likelihood of an event or magnitude occurring.

Curves similar to the one in Figure 3–1 are used both to describe characteristics of a population sample and to predict the probable characteristics of the whole population. This curve represents the line voltage of a circuit as measured on numerous occasions. It is a graph of the magnitude of voltage, plotted on the horizontal or *X* axis, versus the number of times that the magnitude occurred, plotted on the vertical or *Y* axis. As we would expect, voltages near the mean occurred far more frequently than did voltages far from it. The further we go from the mean, the fewer occurrences there are.

The curve in Figure 3–1 is called a *normal distribution curve* because it describes how events occur under normal circumstances. It is also called a probability curve since it describes what would probably occur in a larger group of like incidents. It can be constructed from a multitude of measurements or from a sample of

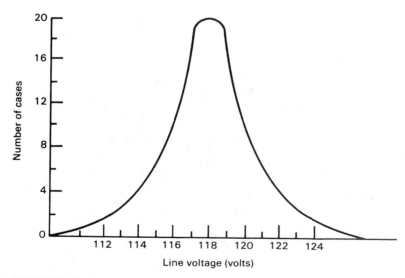

——○ **FIGURE 3–1**
Probability of Line Voltage Values

measurements. We will produce one from our sample data and use it to predict what, if left to chance, the remainder of our resistors will probably be like.

As an aid in our analysis, we use the following rules about probability and the normal curve:

1. Under normal circumstances, events occur in a manner described by a normal, or bell-shaped, curve.
2. Under normal circumstances, the mean, median, midrange, and mode all occur at the same location.
3. 50% of all events occur above the mean and 50% occur below it.
4. 50% of all events occur between -0.67 and $+0.67$ standard deviation from the mean.

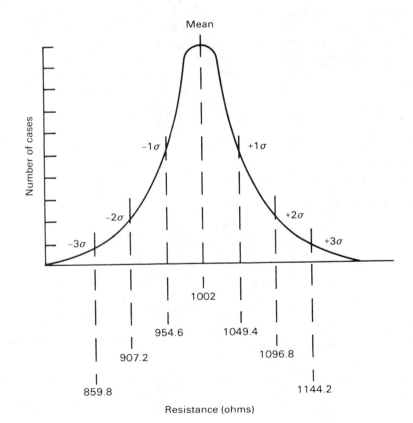

—○ **FIGURE 3–2**
Probability of Resistor Values

5. 68% of all events occur between -1 and $+1$ standard deviation from the mean.
6. 95% of all events occur between -2 and $+2$ standard deviations from the mean.
7. 99% of all events occur between -3 and $+3$ standard deviations from the mean.

○ Example of Using the Normal Curve

Now, we can produce our probability curve and make some predictions. The points we need to plot our curve are the mean and the standard deviation points. The latter are determined by adding to or subtracting from the mean the appropriate multiple of the standard deviation.

$$\overline{X} = 1002\ \Omega \quad \text{and} \quad \sigma = 47.4\ \Omega$$

$$\overline{X} - 1\sigma = 954.6\ \Omega \qquad \overline{X} + 1\sigma = 1049.4\ \Omega$$

$$\overline{X} - 2\sigma = 907.2\ \Omega \qquad \overline{X} + 2\sigma = 1096.8\ \Omega$$

$$\overline{X} - 3\sigma = 859.8\ \Omega \qquad \overline{X} + 3\sigma = 1144.2\ \Omega$$

With this data, we can plot the curve shown in Figure 3–2. It describes what, under normal circumstances, the population that these resistors came from would be like.

From this information, we can predict that there are more resistors with a value of 1002 ohms than with any other value. Also, 68% are between 954.6 and 1049.4 ohms. No more than 5% are less than 907.2 ohms or greater than 1096.8 ohms. There is less than a 1% chance that any resistors are less than 859.8 ohms or greater than 1144.2 ohms.

○ Example of Mean, Median, Mode, Midrange, and Standard Deviation Calculations

Consider the following resistance measurements (in ohms):

1000
991
1002
996
1011
999
1000
998

1004
994
985
1008

Placing these numbers in rank order is the first step in determining the median and the mode. The order from smallest to largest is as follows:

985
991
994
996
998
999
1000
1000
1002
1004
1008
1011

The mode, the number appearing most frequently, is 1000. The median, the number midway through the list, is 999.5 because there is an even number of resistors and 999.5 is midway between the middle pair, 999 and 1000.

Now, we calculate the midrange and mean.

$$\text{midrange} = \frac{\text{largest} + \text{smallest}}{2} = \frac{1011\ \Omega + 985\ \Omega}{2} = 998\ \Omega$$

$$\overline{X} = \frac{\sum x}{n} = \frac{11{,}988\ \Omega}{12} = 999\ \Omega$$

The standard deviation is calculated next. Table 3–5 gives the individual deviations and deviations squared, which are needed for this calculation. From this data, the standard deviation is as follows:

$$\sigma = \sqrt{\frac{\sum d^2}{n-1}} = \sqrt{\frac{556}{11}} = \sqrt{50.5} = 7.1\ \Omega$$

Since $\overline{X} = 999$ ohms, we can state the following:

○ **TABLE 3-5**
Individual Values, Deviations, and Deviations Squared of
Sample Resistors

Value	Deviation from Mean	Deviation Squared
985	−14	196
991	−8	64
994	−5	25
996	−3	9
998	−1	1
999	0	0
1000	+1	1
1000	+1	1
1002	+3	9
1004	+5	25
1008	+9	81
1011	+12	144
		$\Sigma\ d^2\ =\ 556$

$$\overline{X} - 0.67\sigma = 994.2\ \Omega \qquad \overline{X} + 0.67\sigma = 1003.7\ \Omega$$

$$\overline{X} - 1\sigma = 991.9\ \Omega \qquad \overline{X} + 1\sigma = 1006.1\ \Omega$$

$$\overline{X} - 2\sigma = 984.8\ \Omega \qquad \overline{X} + 2\sigma = 1013.2\ \Omega$$

$$\overline{X} - 3\sigma = 977.7\ \Omega \qquad \overline{X} + 3\sigma = 1020.3\ \Omega$$

Thus, resistors randomly selected from the same source as this sample come from will probably fall into the following pattern:

50% will be above 999 ohms and 50% will be below.
50% will be between 994.2 and 1003.7 ohms.
68% will be between 991.9 and 1006.1 ohms.
95% will be between 984.8 and 1013.2 ohms.
99% will be between 997.7 and 1020.3 ohms.

SUMMARY

The method of analysis you can apply to data depends on the scale the data were derived from. Nominal data, such as telephone numbers, use numbers for identification. The only procedure you can perform with this type of data is to count how many different numbers there are. Ordinal scales are the next level, and they indicate only rank order, such as in a list of the top ten songs. Interval scales indicate the distance between points and have an arbitrary zero reference. The highest-level scale is the ratio scale; with it, all data are referenced to a zero that indicates complete absence of the characteristic measured. Voltage, current, and power, for example, are measured on ratio scales. Such data can be subjected to all mathematical calculations.

Many types of errors have been described in this chapter. Instrument, parameter, and observer errors are errors categorized by cause. Systematic and random errors are errors categorized by behavior. Errors should be prevented, when possible, and should be accounted for in data when they are uncontrollable. Limiting error describes the maximum error expected from single or combined sources. Resistor tolerance, for example, describes a limiting error.

Errors in combination are a bit complicated to deal with. When resistors are connected in series, their resistances are added but their individual errors are not. Instead, they are averaged according to their individual weights. Errors in products and quotients are added algebraically. When you are determining limiting error, always assume the worst case. The worst case in products occurs when both errors are in the low limits or both are in the high limits. In quotients, the worst case occurs when one error is in the low limit and the other is in the high limit.

Data are often summarized by calculating a mean, or average. The mean is found by adding all values and dividing the sum by the number of values. While the mean is a useful summary number, other values can also be used to describe central tendency. The median is the central number in a group placed in rank order. The mode is the number that occurs most frequently. The midrange is the value located mathematically between the largest and smallest values.

Standard deviation is used with a mean to describe the center and the variation of a group of measurements. It is the rms of variation and is calculated with a process of squaring, summing, and finding the square root. Standard deviation describes how a total population, represented by a small sample, is likely to be distributed under normal circumstances. According to the rules for probability and the normal curve, 68% of a population will be within ± 1 standard deviation from the mean, 95% within ± 2 standard deviations, and 99% within ± 3 standard deviations. One-half of all values will be within ± 0.67 standard deviation of the mean.

BIBLIOGRAPHY

Electronic measuring instruments; results precise but wrong, J. R. Pearce, Electronics and Power, February, 1983.

Statistics Made Simple, H. T. Hayslett, Jr., Doubleday & Company, Inc., 1968.

Tests and Measurements, 2nd ed., Leona E. Tyler, Prentice-Hall, 1971.

Fundamentals of Physical Measurement, Ernest Zebrowski, Jr., Duxbury Press, 1979.

QUESTIONS

1. Name and compare four levels of measurement.

2. Name the processes that can be performed with numbers at each of the four levels.

3. Which scale is used for measuring voltage and current? In what way is it different from the other three scales?

4. Describe what is meant by the term *significant figures*.

5. How many significant figures are in the number 2.4×10^3? Explain.

6. State how many significant figures are in each of the following numbers: 0.0202, 102, 50.3, 1.0.

7. State how many significant figures are in each of the following numbers: 155.2, 155.20, 0.002, 0.01020.

8. Name five types of error. Give an example of each type of error.

9. Describe how the effects of random error can be reduced.

10. Define and describe *limiting error*.

11. Name and describe four measures of central tendency.

12. How can systematic error be compensated for?

13. Describe the limitations in using the mean for representing central tendency.

14. Why is median income generally used to describe the salary characteristics of a group?

15. What is the resulting tolerance of two 10% resistors connected in series? In parallel?

16. What is the effect, on the calculation of power, of a 3% error in a voltage value and a 4% error in a current measurement?

17. What is the effect, on the calculation of power, of a 3% error in a current value and a 5% error in a resistance value?

18. What is meant by *worst case,* and why should it always be considered?

19. What are the relationships between mean, median, mode, and midrange in a normal distribution?

20. What is meant by *standard deviation,* and why is it important in the description of a group?

21. Give the percentage of a normally distributed group that falls within these values of the mean: $\pm 0.67\sigma$, $\pm 1\sigma$, $\pm 2\sigma$, $\pm 3\sigma$.

PROBLEMS

1. Determine the power in a circuit having a voltage of 40 volts and a current of 0.25 ampere. Determine the limiting error if the voltage is accurate within $\pm 3\%$ and the current is accurate within $\pm 5\%$.

2. Determine the resistance and the limiting error in the circuit described in Problem 1.

3. Determine the power in a circuit having a voltage of 12 volts and a current of 30 milliamperes. Determine the limiting error if the voltage is accurate within $\pm 4\%$ and the current is accurate within $\pm 2\%$.

4. Determine the resistance and the limiting error in the circuit described in Problem 3.

5. Determine the limiting error present when a 1200-ohm, 10% resistor is connected in parallel with a 1500-ohm, 5% resistor.

6. Determine the four measures of central tendency of the following voltages (in volts): 19.0, 20.0, 18.6, 26.0, 26.5, 11.5, 19.1, 18.6, 20.8.

7. Determine the four measures of central tendency of the following voltages (in volts): 117.3, 117.7, 119.0, 118.0, 117.0, 117.8, 118.5, 118.8, 117.7, 117.9, 117.2

8. Determine the standard deviation of the voltages listed in Problem 6. Between what limits will 50% of all voltages probably fall? 95% of all voltages? 99% of all voltages?

9. Determine the standard deviation of the voltages listed in Problem 7. Between what limits will 50% of all voltages probably fall? 95% of all voltages? 99% of all voltages?

10. Draw a graph representing the distribution described in Problems 6 and 8 by plotting the $\pm 0.67\sigma$, $\pm 1\sigma$, $\pm 2\sigma$, and $\pm 3\sigma$ points.

11. Draw a graph representing the distribution described in Problems 7 and 9 by plotting the $\pm 0.67\sigma$, $\pm 1\sigma$, $\pm 2\sigma$, and $\pm 3\sigma$ points.

12. Using the probability table in the Appendix, determine the limits between which 75% of all voltage values described in Problems 6 and 8 will fall.

13. Using the probability table in the Appendix, determine the limits between which 80% of all voltage values described in Problems 7 and 9 will fall.

4

Records and Reports

OBJECTIVES

Documentation is another task of most electronics technicians. This chapter explains the need for documentation, some methods used, and the characteristics of good-quality records. A sample lab report is also included.

Name and explain four reasons for documenting lab activities.
Describe four characteristics of a good-quality document.
Name and describe four types of lab activity records.
Name and describe the parts of a complete lab report.
Prepare a complete report of a lab measurement project.

DOCUMENTATION OF LAB ACTIVITIES

Some sort of documentation is almost always part of a technician's measurement activities. Assignments and procedural instructions are generally written, although they are often written by someone else. A technician usually must report what was done, how it was done, and what the results were. This chapter describes some common methods used for these tasks.

Need for Documentation

Some technicians claim that the only reason for reports is to produce more paperwork. That is far from the truth. One value of records is *organization*. You will plan your activities more carefully if you know that you must eventually prepare a report. You will think twice about equipment selection, measurement procedures, and data collection. Thus, reports can help you organize your activities.

Reports are also used for *communication*. The person who gave you the assignment wants to know your results, usually more detailed than a comment that "it works." People want to see numbers, and they want to know where those numbers came from. Other people, located some distance away, may also need the results. Possibly, one or two years later, you will want to review your activity and your own report. So, documents are needed for communication over both distance and time.

Furthermore, documents produce a *permanent record*. You may not always be in the same job or, in fact, even in the same company. The results of your effort belong to the company; therefore, they are often maintained by the company in a permanent form. Companies may use these records, for example, in patent applications as evidence that work was done on the concept or product in the application.

Finally, military specifications often require *traceability* in the manufacture of components and products. Records show the process of a component from the original manufacturer, to the subassembler, and on to the final system manufacturer. Instrument calibration is often noted on an attached label, stating when the instrument was calibrated and by whom and perhaps giving other information, as needed. The record is thus a guarantee of the instrument's quality.

Characteristics of Quality Documents

The most important characteristic of any record is that it is *accurate*. There will be times when you conduct a series of measurements and the results are not what you anticipated. Your report must represent what happened, not what you wanted or expected to happen. Language must be clear. The method of presentation—that is, your words and figures—must be *understandable* to the reader.

The report also must be *complete*. It should leave no questions about what you did, why and how you did it, and what your results were. There should be enough information in your report for another person to duplicate your activities. These requirements do not necessarily mean that your report must be long. Instead, documents should be *concise,* with as few words as necessary to tell the complete story.

Methods of Documentation

Technicians in experimental labs generally keep some form of bound *notebooks* with numbered pages. Legal requirements often dictate that page entries be in ink, dated, and signed by the technician. Even without such requirements, keeping a bound and dated log is an excellent idea. It prevents the loss of information that is likely to result from loose pages and/or poor memory.

Another method of documentation is the *calibration label.* Attached to a meter, oscilloscope, or other instrument, it indicates calibration date, technician, and other essential data. *Service or repair tags* are attached to defective products, noting problems, action, service person, date, release, and so forth. Service orders and forms provide the same function but are kept in one location rather than with the product.

Reports are prepared to satisfy the needs of the reader. The reader may only need a graph, chart, or table. However, if the report is the only document to be supplied, a more complete *lab report* is prepared. The type of report prepared is dictated by the needs of all persons involved, and the report could include government documents, customer service records, or calibration records.

INDIVIDUAL DOCUMENTS

While reports can be extensive, they can also be simple. Single-page documents such as schematics and graphs are two examples. And while reports may be part of a larger document, they should be able to stand alone. That is, even single-page reports should be titled, dated, and self-explanatory, with the author's name included. The following subsections describe some simple reports technicians must often prepare.

Schematics and Block Diagrams

Once again, primary considerations in schematics and block diagrams are accuracy and completeness. Make sure that all information needed for reader understanding is included and that it is correct. Component values and any special test conditions should also be included.

Schematics show specific components in a product. The VOM schematic in Figure 4–1 is a typical report. Notice that most component values are given. Special parts such as switches and diodes are identified on a separate parts list.

Block diagrams show the flow of information for the product.

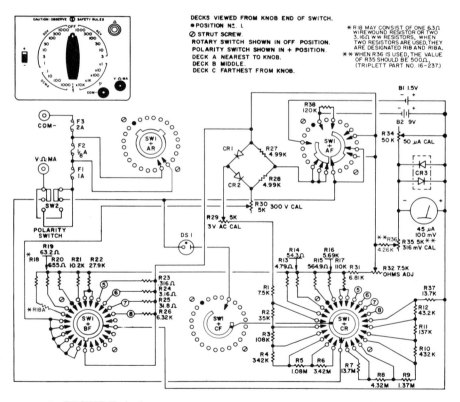

───○ **FIGURE 4–1**
Schematic for a VOM Circuit (Reproduced with permission from Triplett Corporation)

For example, Figure 4–2 shows the sections of a power supply. Here, the first block represents one component, and the fourth block represents many components. Notice also that information flows from left to right.

Block diagrams can also show how one product or item of equipment is connected to others. This type of report is useful for explaining how you conducted an experiment. Figure 4–3 is an example of this type of block diagram.

Tables

Tables provide an orderly way of presenting data. Carefully selected column headings are needed if a table is to stand alone. The number of columns used depends on the purpose of the measurements and

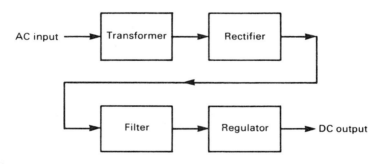

──○ **FIGURE 4–2**
Block Diagram of a Power Supply

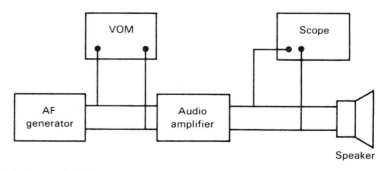

──○ **FIGURE 4–3**
Block Diagram of an Experimental Configuration

the characteristics of the data. For example, consider the data in Table 4–1. It shows the output voltages of a power supply for various control settings. The *independent variable*—what is manipulated— is located in the left column of the table. The *dependent variable*— what depends on or is influenced by the independent variable—is in the right column of the table.

Sometimes, the addition of another column improves the communication process. Table 4–2 summarizes calculations and measurements that compare the color-coded and measured values of some resistors. All a reader needs to know about these results is

TABLE 4–1

Output Voltages of a Power Supply for Various Control Settings

Dial Setting	Output Voltage (Volts DC)
0.0	0.0
1.0	1.8
2.0	3.5
3.0	5.0
4.0	7.0
5.0	8.6
6.0	10.0
7.0	12.0
8.0	13.4
8.6	14.0
9.0	14.3
9.4	14.5
10.0	14.6

TABLE 4–2

Color-Coded and Measured Values of Resistors

Resistor	Coded Value (Ohms)	Measured Value (Ohms)
R_1	1000	947
R_2	1000	1023
R_3	470	478
R_4	2200	2095
R_5	330	345
R_6	4700	4640

presented in this one table. Additional information about procedures can be noted elsewhere.

Graphs

Most *graphs* are produced by plotting points that represent the magnitude of one parameter as a function of another. If enough points are plotted, the points will eventually create a line or curve. However, that situation does not normally arise. Typically, a dozen or so points are plotted. The location of the values in between are then estimated. Because they summarize data so well, graphs are generally more effective in communicating than tables, although they are less precise.

Figure 4–4 shows a simple graph of the data presented in Table 4–1. The independent variable is the dial setting and is described on

——○ **FIGURE 4–4**
Graph of Power Supply Output Voltages for Various Control Settings

the horizontal (X) axis. The dependent variable is output voltage and is described on the vertical (Y) axis. Each axis description usually has three parts: parameter, units, and magnitude. On the vertical axis, these elements are output voltage, volts, and the appropriate numbers, respectively. In this example, the horizontal axis has only two parts, the dial setting and the numbers. There are no units for the dial setting.

The curve in Figure 4–4, rather than being drawn directly from point to point, is drawn as a smooth line that best fits the plotted points. The points were plotted first. Then, the curve was drawn with a French curve, a device shown in Figure 4–5. Graphs are generally drawn in this way. It is assumed that the function is, in fact, a smooth function and that any variations in it are caused by measurement error. The specific measured values are recorded, as shown in Figure 4–4, by plotting them and placing small circles around them.

Figure 4–6 shows the frequency response characteristics of an audio amplifier. This curve represents gain, in decibels, as a function of frequency, in hertz. Since the vertical axis of the graph is linear (equally spaced), while the horizontal axis is logarithmic, Figure 4–6 is a *semilogarithmic graph*. It is a four-cycle graph since the logarithmic pattern appears four times. The left end represents 10

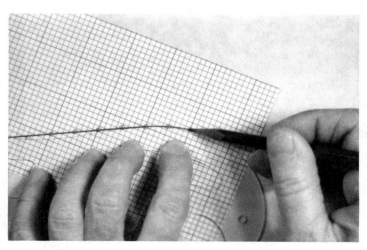

──○ **FIGURE 4–5**
French Curve Used for Drawing Curved Lines

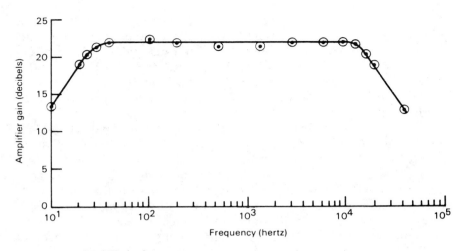

────○ **FIGURE 4–6**
Semilogarithmic Graph of Audio Amplifier Frequency Response

hertz, and each cycle that follows is one decade, or ten times, higher than the cycle before it.

Figure 4–7 shows a *polar coordinate graph* of the relative field strength of an antenna as a function of horizontal direction. The independent variable is the angle of horizontal direction, in degrees, and the dependent variable is relative field strength, in decibels. Power numbers are not given since they depend on transmitter power and distance. Persons using this graph would be interested in the relative field strength in front of the antenna as compared with the strength at points to each side.

In summary, there are five points to remember when you are preparing a graph:

1. Locate the independent variable on the X axis and the dependent variable on the Y axis.
2. Indicate the parameter, units, and numbers for each axis.
3. Plot enough points to show the general shape of the curve. Plot extra points in areas of unusual change.
4. Draw a curve that represents the best fit of the points plotted.
5. Add a title, date, your name, and any other information that will help the graph stand alone.

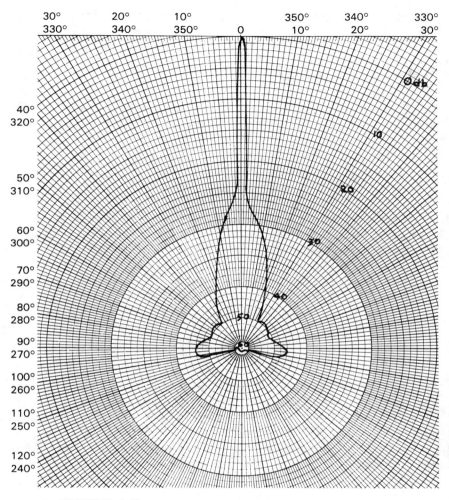

——○ **FIGURE 4–7**
Polar Coordinate Graph of Microwave Antenna Pattern

LAB REPORTS

Quite often, a single graph or table is not enough information for
the person requesting the measurement. More detail is often wanted.
A *lab report* is a comprehensive, after-the-fact statement about what
you intended to do, what you did, and what your results and con-

clusions were. It should leave no unanswered questions with the reader. This section describes the major parts of a lab report.

Preliminary Section

The first part of a lab report is a *title page,* which has a title, your name, and the date of the study. Examples of titles are as follows:

> Power Line Voltage Measurements
> Frequency Response of an Audio Amplifier
> Standing Wave Ratio of an Antenna System

The first item on the next page of the report is a statement of your *objectives.* The objectives describe the purpose of your measurements. Sample objective statements are listed next.

> Determine the mean, the range, and the standard deviation of a line voltage during a one-week period.
> Determine the frequency response of an audio amplifier under 25% of maximum load.
> Determine the standing-wave ratio of a selected antenna system, using 1/8-, 1/4-, and 1/2-wave dipole antennas.

The last part of your intentions section involves *equipment and materials.* Here, you list specific test equipment by brand name, equipment type (VOM, for example), model number, and serial number. This section documents the qualifications of the instruments used. The materials list includes the components and other products used to fabricate the product or object of the measurements.

Procedures Section

The first item in this section is called a *procedures statement,* and it briefly presents what you did in your measurement. It is a general overview used in conjunction with illustrations to explain how the study was carried out. A sample procedures paragraph follows (the figure and table are not included here).

> An audio amplifier circuit was connected according to the diagram in Figure A. The input voltage was adjusted for 0.25 volt rms. The input

frequency was varied from 10 to 30,000 hertz, and the output voltage was measured and recorded (see Table A). Output power as a function of frequency was calculated and plotted.

Other effective parts of the procedures section are *schematics* and *block diagrams*. They can eliminate many words. A schematic can specifically describe how components are connected to produce a circuit. Block diagrams can indicate how test instruments and other preassembled items are connected to each other.

Finally, *tables* should be used in this section for the recording of data. (They also serve to remind you of the data you need.) Tables keep data organized and help readers find it and interpret it.

The combination of procedures statement, schematic, block diagram, and data tables provides a brief yet complete explanation of what was done in a measurement project.

Results/Conclusions Section

Once your measurements are complete, you will probably perform some calculations. In the final section of your report, *sample calculations* show the reader how you processed your data. For example, you may have measured the output voltage of an audio amplifier and calculated the gain. One sample calculation showing the formula you used and how you used it is appropriate. All other calculated gain values can simply be recorded in a table. One typical application of each formula used in your study should be part of your report.

Graphs or tables can be used in this section, also, to summarize your results. For instance, a semilogarithmic graph is an appropriate way to report the results of frequency response measurements. Tables are useful for comparing multiple variables. For example, you might use a table to report the output voltage of a power supply as a function of dial setting under varying load conditions. Columns could be added for other factors such as regulation percentage or load resistance. You will have to decide which method of reporting results is the most effective for each study you do.

The last part of a report is called *conclusions*, and it is unlike the other sections. Up to now, all information was factual. The conclusions statement, however, is a subjective summary—your opinion of the data in light of the objective of the investigation. That is, you had an objective and you made some measurements. What did the measurements tell you? The answer to this question leads to your conclusions.

SAMPLE LAB REPORT

The content and length of lab reports vary with the objectives. A report should only be as long as it needs to be to achieve the objectives. The first page is usually a cover, which has a title, your name, and the date of the study. Your reports should be somewhat similar to the sample report shown in Figure 4–8, which appears on pages 66–67. (The cover is not included in this sample.)

SUMMARY

Records are necessary for almost all laboratory activities. Government documents, customer service orders, calibration charts, and lab reports are just a few examples of required paperwork. They provide effective methods for communicating what needs to be done and what was done. Good-quality documents are accurate, clear, complete, and concise, yet leave no unanswered questions. Documents can be as simple as a graph, a table, or a tag or as complex as an extensive formal report.

Schematics and block diagrams are useful in describing the internal connections of a product or the external interconnection of products and instruments. Both eliminate many words from a report. Tables organize and present data in an easily understood summary form. They are especially useful in multiple-variable situations. Graphs present a more vivid picture of how one parameter responds as a function of others. Graphs, however, lack the accuracy and precision of tables.

A lab report is a comprehensive summary of laboratory activities or processes. It describes the intent, procedures, and results of measurement or other lab efforts. The parts of a report are as follows: the preliminary section (title, name, date, objective, equipment, and materials); the procedures section (procedures statement, schematics, block diagrams, and tables); and the results/conclusions section (sample calculations, graphs, tables, and conclusions).

A document must reflect the objective of the assignment. It should be accurate, clear, and concise. It should indicate what *did* happen, not what you hoped or expected to happen. Finally, a document should not leave a reader with any questions about what you did. It must be complete and self-explanatory.

BIBLIOGRAPHY

Handbook of Mathematical Functions with Formulas, Graphs, and Mathematical Tables, 5th ed., Richard S. Burington, ed., McGraw-Hill Publishing Company, 1973.

Fundamentals of Physical Measurement, Ernest Zebrowski, Jr., Duxbury Press, 1979.

Writing for Technicians, Marva T. Barnett, Delmar, 1982.

QUESTIONS

1. Name and explain four reasons for documenting lab activities.

2. Name and describe four ways of documenting lab activities.

3. Explain why lab notes are often kept in bound books with numbered pages, are written in ink, and are signed and dated.

4. Describe the characteristics of good-quality lab records.

5. Explain what is meant by the phrase "communicating over distance and time."

6. Compare the use of schematics with the use of block diagrams.

7. Compare the qualities of graphs and tables.

8. Name and describe the major parts of a graph.

9. Compare linear, logarithmic, and polar coordinate graphs.

10. Explain what is meant by *best-fit curve*. Why is it used?

11. Explain the difference between dependent and independent variables. On which axis of a graph is each plotted?

12. Where are the dependent and independent variables located in a table?

13. Name and describe the major parts of a complete lab report.

14. Explain why the equipment used should be identified by brand, model, and serial number.

15. Describe the characteristics of a procedures statement.

16. Discuss the relationship between objectives and conclusions statements.

17. Explain what is meant by "fact versus opinion" in a lab report. Indicate sections that contain facts and sections that contain opinion.

OBJECTIVE

Determine the effects of load upon the output voltage
and ripple of a XXXX model 925 DC power supply. Calculate the
regulation and ripple percentage.

EQUIPMENT and MATERIALS

XXXX Model 925 DC power supply, serial number 326
Fluke model 8050A digital multimeter
Leader model LBO-517 oscilloscope, serial number 3102
Five 10-watt load resistors

PROCEDURE

The equipment and materials were assembled, as shown
in Figure A. The DC output voltage was measured with the
multimeter and recorded in Table A. The AC voltage was measured
with the oscilloscope and also recorded in the table. The
load resistance was varied and the measurements repeated.
Regulation percentage and ripple percentage were calculated
for five output levels and were recorded in Table A. Power
supply output characteristics were graphed in Figure B.

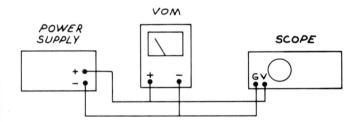

FIGURE A. POWER SUPPLY TEST CONNECTIONS

TABLE A. LOAD, OUTPUT, REGULATION, AND RIPPLE

| Load | Output | | | Regulation | Ripple |
(Ohms)	Volts DC	Volts pp	Volts rms	(Percent)	(Percent)
Open	10.0	0.00	0.00	0.0	0.0
400	10.0	0.00	0.00	0.0	0.0
300	10.0	0.00	0.00	0.0	0.0
200	10.0	0.14	0.05	0.0	0.5
100	9.4	0.39	0.14	6.4	1.4
20	8.4	0.73	0.26	19.0	3.0

──────○ **FIGURE 4–8**
Sample Lab Report

SAMPLE CALCULATIONS

$$\text{regulation} = \frac{(V_{nl} - V_{fl})}{V_{fl}} \times 100\%$$

$$= \frac{(10 \text{ V} - 8.4 \text{ V})}{8.4 \text{ V}} \times 100\% = 19\%$$

$$\text{ripple percentage} = \frac{V_{rms}}{V_{DC}} \times 100\%$$

$$= \frac{0.26 \text{ V}}{8.4 \text{ V}} \times 100\% = 3\%$$

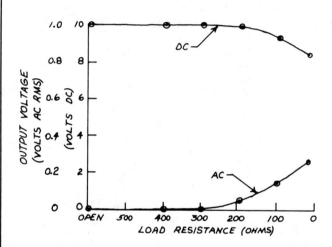

FIGURE B. POWER SUPPLY OUTPUT CHARACTERISTICS

CONCLUSIONS

The DC output voltage of this power supply varied from 10 volts with no load to a low of 8.4 volts under the specified full load. Anyone using this power supply would have to measure and readjust the output voltage after connecting the load. No AC ripple was detected under no-load conditions. However, the ripple increased to a level of 0.26 volt rms at maximum load. This load produced a ripple level of 3%, which could be a concern in some applications.

PROBLEMS

1. Plot a graph of the following power supply test results.

Load Current (Milliamperes)	Ripple (Percent)
0	0.100
50	0.130
100	0.180
150	0.220
200	0.280
250	0.340
300	0.340
350	0.360
400	0.400

2. Plot a graph of the following voltmeter calibration data.

True Voltage (Volts)	Indicated Value (Volts)
0.00	0.00
1.00	1.19
2.00	2.48
3.00	3.59
4.00	4.57
5.00	5.52
6.00	6.41
7.00	7.20
8.00	8.00
9.00	8.70
10.00	9.42

3. Plot a graph of the following circuit frequency response data. The input at all frequencies is 50 millivolts.

Frequency (Hertz)	Output (Volts)
20	0.281
30	0.446
45	0.655
80	0.671
200	0.670
500	0.671
1500	0.671
3000	0.650
4800	0.445
7000	0.279

4. Plot a graph of the following circuit frequency response data. The input at all frequencies is 100 millivolts.

Frequency (Hertz)	Output (Volts)
350	0.134
500	0.188
700	0.260
900	0.299
1,400	0.316
3,000	0.315
10,000	0.316
20,000	0.314
37,000	0.282
50,000	0.237
70,000	0.168
100,000	0.102

5

Analog Meters

OBJECTIVES

For years, analog meters were the most common measurement instruments in electronics. Today, they share that popularity with digital meters. Although digital meters become more common each day, their analog counterparts remain popular. Analog meters are popular because they are the foundation of many measurement instruments. This chapter describes basic analog meters and their circuits, specifications, and applications.

> Describe the operation of a meter movement with taut-band suspension.
>
> Determine the multiplier and shunt values, true accuracy, input resistance, and loading effect of a multimeter.
>
> Name and describe four common specifications of an analog multimeter.
>
> Describe the operational characteristics of an electronic multimeter, a thermocouple meter, and an electrodynamometer.
>
> Measure voltage, current, resistance, and decibels with an analog multimeter.

ANALOG MULTIMETER THEORY

There are probably more analog multimeters in electronics labs today than any other type of instrument. Figure 5–1 shows such an instrument. They are called *analog* because their pointer deflection

──○ **FIGURE 5–1**
Volt-Ohm Milliammeter (Photo courtesy of Simpson Electric
Company)

along their calibrated scale is analogous, or proportional, to the magnitude of the parameter being measured. The parameters are usually
voltage, resistance, or current, and for that reason, the instruments
are called volt-ohm milliammeters (VOMs). Behind each scale is the
heart of the instrument, a meter movement. In most cases, it is an
electromagnetic movement with a taut-band suspension. Our discussion of analog meters begins here with the movement and scale.

Movements and Scales

The permanent-magnet, moving-coil (PMMC) meter movement consists of a coil suspended within a permanent magnet, as shown in
Figure 5–2A. Current through the coil produces a magnetic field,
which reacts with the permanent magnetic field, causing the coil to
rotate. This rotation, in turn, moves a pointer up the scale a distance
proportional to the amount of coil current. Reversing the current

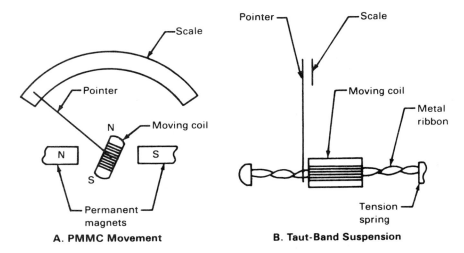

A. PMMC Movement

B. Taut-Band Suspension

C. Jeweled-Bearing Suspension

○ **FIGURE 5-2**
Analog Multimeter

direction causes the pointer to move down the scale and rest against
a mechanical stop.

A key factor in the performance of a meter is the ease of coil
rotation. The taut-band suspension system, shown in Figure 5–2B,
supports the coil and allows it to rotate with very little friction. Its
name is derived from the coil support, a fine metal ribbon that has
been twisted and is held taut with tension springs. The jeweled-
bearing system in Figure 5-2C utilizes the fine points of an axle that

rotates within jeweled bearings. At each end of the coil is a spiral spring with one end attached to the coil and the other attached to a fixed position. These springs carry coil current between the coil and stationary points and keep the pointer at zero when no current is flowing. The taut-band system is less prone to friction from wear than the jeweled-bearing system and more tolerant of rough handling or being dropped. For these reasons, it is becoming the standard of the industry.

A single-range PMMC meter generally has a linear (evenly spaced) scale with zero at the left end. Sometimes, zero is located at the center of the scale for indicating parameters, such as frequency deviation around some predetermined reference point. The typical meter uses a multirange scale similar to the one in Figure 5–3.

In the scale in Figure 5–3, the numbers 10, 50, and 250 represent the full-scale values for AC voltage, DC voltage, and DC current. The scales above are for DC measurements, and the scales below are for AC. A special scale is used for the 2.5-volt AC range because of rectifier nonlinearity at low voltages (a topic to be described shortly). This meter, like most meters, has one ohms scale; it is on the very top and has zero on the right. The bottom scale is used for measuring voltage or current ratio in decibels. Table 5–1 shows the values indicated by the pointer and scales for various meter ranges.

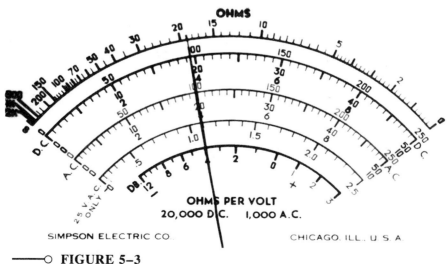

FIGURE 5–3
Multimeter Scale

○ **TABLE 5–1**
Values Indicated by Pointer and Scales for Various Meter Ranges

Range (R × 1)	Indication (19 Ω)
250 V DC	97 V
250 V AC	100 V
50 V DC	19.5 V
50 V AC	20 V
10 V DC	3.9 V
10 V AC	4 V
2.5 V DC	0.97 V
2.5 V AC	1.1 V

Parallax error is an error in the interpretation or reading of a pointer and scale indication caused by the misalignment of the viewer from the intended position. For example, an automobile driver and a passenger will probably read different speeds from a speedometer because of their different alignments with pointer and scale. Parallax error is reduced in meters by the use of knife-edge pointers and mirrors on the scales. Proper viewing occurs when the viewer aligns the pointer with its reflection.

Voltmeter Circuits

As indicated earlier, current flowing through a meter coil causes upscale pointer deflection. The amount of voltage this coil needs for full-scale current depends on the coil's resistance. If we know the coil's resistance and the full-scale current, we can calculate full-scale voltage and use the meter directly as a single-range voltmeter.

○ **Example of Calculating Full-Scale Voltage Needed**

Consider a coil with a resistance of 5000 ohms that requires 50 microamperes of current for full-scale deflection. The voltage needed is as follows:

$$V = I \times R = 0.00005 \, A \times 5000 \, \Omega = 0.25 \, V$$

This meter can be used as either a 0.25-volt or a 50-microampere, *full-scale deflection* (FSD) *meter,* with no modification. That is, its pointer will move to the highest end of the scale when 0.25 volt is placed across it or 50 microamperes is passed through its coil. Since it has a coil resistance of 5000 ohms, one action will cause the other.

Whenever a different range is required, a circuit like the one in Figure 5–4 is often designed. In this case, a 1-volt FSD meter is desired. This voltage range is produced by adding resistor R_1 to the circuit; R_1 is a series resistor called a *multiplier.* Its resistance is such that it and the inherent meter resistance R_m limit the meter current to the amount necessary for full-scale deflection when the desired FSD voltage is applied.

○ **Example of Analyzing a Voltmeter Circuit**

The circuit in Figure 5–4 can be analyzed as a simple series circuit. The resistances are as follows:

$$R_T = \frac{V}{I} = \frac{1\ V}{50\ \mu A} = 20{,}000\ \Omega$$

$$R_T = R_m + R_1$$

$$R_1 = R_T - R_m = 20{,}000\ \Omega - 5000\ \Omega = 15{,}000\ \Omega$$

Resistor R_1 is 15,000 ohms because it and the 5000-ohm coil limit the current to 50 microamperes when 1 volt is applied to the meter input.

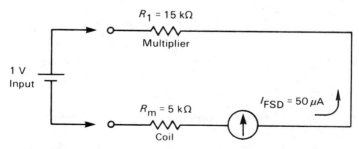

—○ **FIGURE 5–4**
Basic Voltmeter Circuit

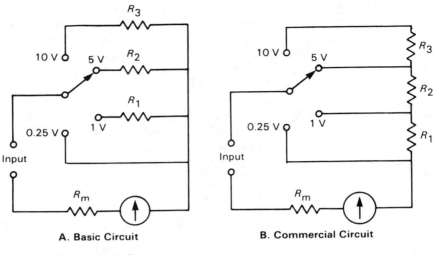

——○ FIGURE 5–5
Multirange Voltmeter Circuits

Most voltmeters have a number of ranges, as indicated in Figure 5–5. Multiplier values are determined in the same manner they were determined in the previous example.

○ Example of Calculations for the Basic Multirange Voltmeter Circuit

Refer to the basic voltmeter circuit in Figure 5–5A. The multiplier for the 5-volt range is R_2, and its value is determined as follows:

$$R_T = \frac{V}{I} = \frac{5\,V}{50\,\mu A} = 100{,}000\,\Omega$$

$$R_T = R_m + R_2$$

$$R_2 = R_T - R_m = 100{,}000\,\Omega - 5000\,\Omega = 95{,}000\,\Omega$$

The multiplier for the 10-volt range is R_3, and its value is determined in this way:

$$R_T = \frac{V}{I} = \frac{10 \text{ V}}{50 \text{ }\mu\text{A}} = 200,000 \text{ }\Omega$$

$$R_T = R_m + R_3$$

$$R_3 = R_T - R_m = 200,000 \text{ }\Omega - 5000 \text{ }\Omega = 195,000 \text{ }\Omega$$

○ Example of Calculations for the Commercial Voltmeter Circuit

Calculations for the commercial version of the multirange voltmeter, shown in Figure 5–5B, are as follows. For the 0.25-volt range, R_T is calculated as shown next. No multiplier is required here.

$$R_T = \frac{V}{I} = \frac{0.25 \text{ V}}{50 \text{ }\mu\text{A}} = 5000 \text{ }\Omega$$

For the 1-volt range, the resistances are as follows:

$$R_T = \frac{V}{I} = \frac{1 \text{ V}}{50 \text{ }\mu\text{A}} = 20,000 \text{ }\Omega$$

$$R_1 = R_T - R_m = 20,000 \text{ }\Omega - 5000 \text{ }\Omega = 15,000 \text{ }\Omega$$

For the 5-volt range, the resistances are as follows:

$$R_T = \frac{V}{I} = \frac{5 \text{ V}}{50 \text{ }\mu\text{A}} = 100,000 \text{ }\Omega$$

$$R_2 = R_T - R_1 - R_m = 100,000 \text{ }\Omega - 15,000 \text{ }\Omega - 5000 \text{ }\Omega$$
$$= 80,000 \text{ }\Omega$$

For the 10-volt range, the resistances are as follows:

$$R_T = \frac{V}{I} = \frac{10 \text{ V}}{50 \text{ }\mu\text{A}} = 200,000 \text{ }\Omega$$

$$R_3 = R_T - R_1 - R_2 - R_m$$
$$= 200,000 \text{ }\Omega - 15,000 \text{ }\Omega - 80,000 \text{ }\Omega - 5000 \text{ }\Omega = 100,000 \text{ }\Omega$$

Because multiplier resistors control meter accuracy, stable precision resistors are used in multirange voltmeters.

As mentioned earlier, the voltmeter pointer moves upscale for one polarity and downscale for the opposite polarity. Since alternating current involves continually changing polarities, the meter needs modification before AC can be measured. When the AC voltage mode is selected, instrument rectifiers are switched in. Then, the meter averages the pulsating value.

Figure 5–6 shows two common meter rectifier circuits. The circuit in Figure 5–6A is a full-wave bridge. It converts the incoming AC signal to pulsating DC, which can then be averaged by the coil. Figure 5–6B shows a commonly used half-wave circuit. In this circuit, diode D_1 is the typical half-wave rectifier, and it conducts during one-half of the input cycle. Diode D_2 is off at that time. During the second half of the input cycle, D_1 is off and D_2 is on, causing

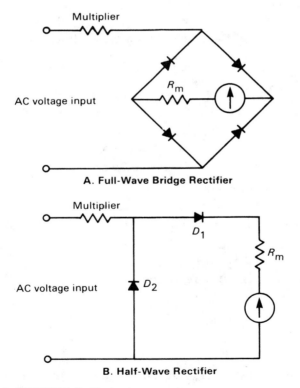

A. Full-Wave Bridge Rectifier

B. Half-Wave Rectifier

FIGURE 5–6
Voltmeter Rectifiers

conduction that bypasses the coil. In this way, D_1 is prevented from having any reverse leakage current, which would increase the average flow and create error.

The half-cycle average of a perfect sine wave equals 0.637 times its peak value. This value is the one that the voltmeter responds to. Since it is the rms or effective value that we want, the meter is calibrated to indicate rms. A signal with a peak value of 1 volt would be seen by the meter as having an average value of 0.637 volt, but the pointer and scale indication would be 0.707 volt. This value is correct and is acceptable for pure sine waves. However, significant error occurs with other waveshapes. More will be said about this topic shortly.

Ammeter Circuits

The basic meter movement described earlier can be used as a 50-microampere FSD ammeter just as it is. As with the voltmeter, it can also be modified to have more current ranges. This modification is done by the addition of parallel resistors, called *shunts*, that provide alternative paths for current beyond the coil range. The ammeter in Figure 5–7 has had its FSD current range extended from 50 to 200 microamperes through the use of shunts.

The total current to be measured (I_T) enters and exits the meter terminals. The meter FSD current (I_m) goes through the coil, while a larger proportional amount (I_s) passes through the shunt

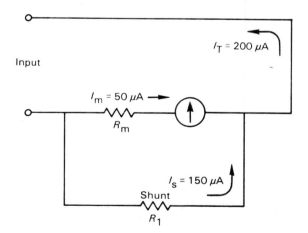

Input

$I_T = 200\ \mu A$

$I_m = 50\ \mu A$

R_m

$I_s = 150\ \mu A$

Shunt

R_1

FIGURE 5–7
Basic Ammeter Circuit

resistor R_1. The shunt value can be determined with simple parallel circuit analysis.

○ Example of Calculations for the Basic Ammeter Circuit

The calculations for the basic ammeter circuit shown in Figure 5–7 are as follows:

$$V_m = I_m \times R_m = 50\,\mu A \times 5000\,\Omega = 0.25\,V$$

$$I_s = I_T - I_m = 200\,\mu A - 50\,\mu A = 150\,\mu A$$

$$R_1 = \frac{V_m}{I_s} = \frac{0.25\,V}{150\,\mu A} = 1666.7\,\Omega$$

If 200 microamperes pass through this meter, 50 microamperes go through the coil, causing full-scale deflection. The remaining 150 microamperes go through the shunt.

Like a voltmeter, an ammeter can also be modified to provide more ranges. In a multirange ammeter, more shunts are added to the basic circuit, as shown in Figure 5–8. In this circuit, no shunt is used for the 50-microampere range since that amount of current

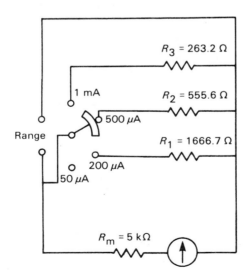

───○ **FIGURE 5–8**
Multirange Ammeter Circuit

causes exactly full-scale deflection. The 200-microampere range was just described. For measurement of that amount, 150 microamperes must be shunted around the coil. The 1667-ohm resistor provides that path. Calculations for the other ranges are similar to the calculations shown in the previous example.

○ **Example of Calculations for the Multirange Ammeter Circuit**

For the 500-microampere range, R_2 is determined as follows:

$$V_m = I_m \times R_m = 50\ \mu A \times 5000\ \Omega = 0.25\ V$$

$$I_2 = I_T - I_m = 500\ \mu A - 50\ \mu A = 450\ \mu A$$

$$R_2 = \frac{V_m}{I_2} = \frac{0.25\ V}{450\ \mu A} = 555.6\ \Omega$$

For the 1-milliampere range, R_3 is calculated in this way:

$$I_3 = I_T - I_m = 1\ mA - 50\ \mu A = 950\ \mu A$$

$$R_3 = \frac{V_m}{I_3} = \frac{0.25\ V}{950\ \mu A} = 263.2\ \Omega$$

As you can see by the progression, shunt resistance decreases and shunt current increases as the ranges get higher. Low-resistance, wire-wound resistors are used to tolerate the high power while maintaining a constant resistance value. A make-before-break switch is used to ensure that the meter is never without a shunt.

The *Ayrton shunt* circuit provides additional protection for a meter by ensuring that a shunt is always connected across the coil. Figure 5–9 shows the circuit, which is somewhat different from the circuit in Figure 5–8. So, also, is its analysis. On the lowest range of 0.1 milliampere, the three shunt resistors form a series group in parallel with the coil. On the middle range of 0.5 milliampere, R_1 and R_2 form a series pair in parallel with the series pair of R_3 and R_m. On the highest range of 1 milliampere, R_1 is in parallel with the series trio of R_2, R_3, and R_m.

○ **Example of Calculations for the Ayrton Shunt Circuit**

Circuit design for the Ayrton shunt circuit gets a bit complicated because there are three unknown resistances with interdependent values. With three unknowns, three equations are required.

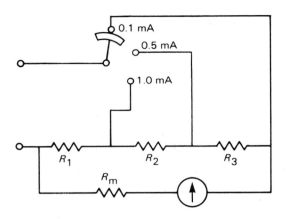

──o **FIGURE 5–9**
Ayrton Shunt Ammeter Circuit

The procedure for the 100-microampere range is as follows:

$$I_s = I_T - I_m = 100\ \mu A - 50\ \mu A = 50\ \mu A$$

Since both paths have equal current, they must have equal resistance. Therefore, the equation for R_m is as follows:

$$R_1 + R_2 + R_3 = R_m \tag{1}$$

For the 500-microampere range, the equations are as follows:

$$I_s = I_T - I_m = 500\ \mu A - 50\ \mu A = 450\ \mu A$$

or

$$I_s = 9I_m$$

Because I_s passes through $R_1 + R_2$, because I_m passes through $R_3 + R_m$, and because of the inverse relationship between circuit current and circuit resistance, we also have the following equation:

$$R_1 + R_2 = \frac{R_3 + R_m}{9}$$

or

$$9R_1 + 9R_2 = R_3 + R_m$$

This equation can also be written as follows:

$$9R_1 + 9R_2 - R_3 = R_m \tag{2}$$

For the 1-milliampere range, the analysis is as follows:

$$I_s = I_T - I_m = 1 \text{ mA} - 50 \text{ μA} = 950 \text{ μA}$$

or

$$I_s = 19I_m$$

Considering the relationship of resistance and the paths of the currents, we can write the following equation:

$$R_1 = \frac{R_2 + R_3 + R_m}{19}$$

or

$$19R_1 = R_2 + R_3 + R_m$$

We can write this equation another way:

$$19R_1 - R_2 - R_3 = R_m \tag{3}$$

Now, our three equations are as follows:

$$R_1 + R_2 + R_3 = R_m$$

$$9R_1 + 9R_2 - R_3 = R_m$$

$$19R_1 - R_2 - R_3 = R_m$$

Through algebraic techniques, we can solve these equations for R_1, R_2, and R_3 (R_m is known). Let us add Equation (1) to Equation (3):

$$
\begin{array}{rl}
19R_1 - R_2 - R_3 = & R_m \\
+\ R_1 + R_2 + R_3 = & R_m \\
\hline
20R_1 \qquad\qquad = & 2R_m
\end{array}
$$

or

$$10R_1 = R_m$$

Since $R_m = 5000$ ohms, $R_1 = 500$ ohms. Next, we add Equation (1) to Equation (2):

$$
\begin{array}{rl}
9R_1 + 9R_2 - R_3 = & R_m \\
+\ R_1 + R_2 + R_3 = & R_m \\
\hline
10R_1 + 10R_2 \qquad = & 2R_m
\end{array}
$$

or

$$10R_2 = 10{,}000 \ \Omega - 5000 \ \Omega = 5000 \ \Omega$$

$$R_2 = 500 \ \Omega$$

Finally, we substitute these values into Equation (1) after we solve Equation (1) for R_3:

$$R_1 + R_2 + R_3 = R_m$$

$$R_3 = R_m - R_1 - R_2 = 5000\,\Omega - 500\,\Omega - 500\,\Omega = 4000\,\Omega$$

Ohmmeter Circuits

Resistance is determined by measuring the current that flows when the unknown resistance is connected to a circuit energized by a battery. Two circuits are generally used, series and shunt. Figure 5–10 shows the basic arrangement of a series ohmmeter. For resistance measurements, current flows through the unknown resistance R_x, the range resistance R_1, and the meter and its resistance R_m, causing pointer deflection. The amount of deflection depends on the amount of current flowing or the inverse of the resistance present. An adjustable resistor R_z is connected in parallel with the meter and serves as a *zero control*.

○ **Example of Calculations for the Series Ohmmeter Circuit**

Circuit calculations when the input leads are shorted ($R_x = 0$) are as follows:

$$I_T = \frac{V_T}{R_T} = \frac{1.5\,\text{V}}{0\,\Omega + 12.5\,\text{k}\Omega + 2.5\,\text{k}\Omega} = 100\,\mu\text{A}$$

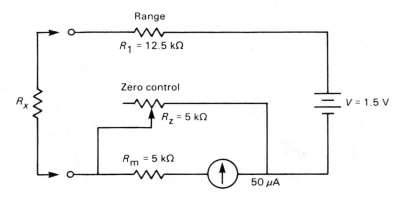

──○ **FIGURE 5–10**
Series Ohmmeter Circuit

Since R_z and R_m are equal and in parallel, the total current I_T will divide, and the meter will have its FSD current of 50 microamperes.

If full-scale deflection does not occur because of factors such as the battery not being exactly 1.5 volts, the zero control R_z is adjusted to change its shunting effect accordingly. The amount of unknown resistance that would cause 1/2-scale deflection can be calculated as follows:

$$I_T = \frac{V_T}{R_T} = \frac{V_T}{R_x + R_1 + [(R_2 \times R_m)/(R_2 + R_m)]}$$

$$R_x = \frac{V_T}{I_T} - R_1 - \frac{R_2 \times R_m}{R_2 + R_m} = \frac{1.5 \text{ V}}{0.00005 \text{ A}} - 12{,}500 \text{ }\Omega - 2500 \text{ }\Omega$$

$$= 30{,}000 \text{ }\Omega - 12{,}500 \text{ }\Omega - 2500 \text{ }\Omega = 15{,}000 \text{ }\Omega$$

For 1/4-scale deflection, the unknown resistance is as follows:

$$R_x = \frac{V_T}{I_T} - R_1 - \frac{R_2 \times R_m}{R_2 + R_m} = \frac{1.5 \text{ V}}{0.000025 \text{ A}} - 12{,}500 \text{ }\Omega - 2500 \text{ }\Omega$$

$$= 60{,}000 \text{ }\Omega - 12{,}500 \text{ }\Omega - 2500 \text{ }\Omega = 45{,}000 \text{ }\Omega$$

For 3/4-scale deflection, the unknown resistance is as follows:

$$R_x = \frac{V_T}{I_T} - R_1 - \frac{R_2 \times R_m}{R_2 + R_m} = \frac{1.5 \text{ V}}{0.000075 \text{ A}} - 12{,}500 \text{ }\Omega - 2500 \text{ }\Omega$$

$$= 20{,}000 \text{ }\Omega - 12{,}500 \text{ }\Omega - 2500 \text{ }\Omega = 5000 \text{ }\Omega$$

There is no current—and, therefore, no deflection—when no R_x is connected to the terminals.

Figure 5–11A shows a series ohmmeter scale, which has zero on the right end and is nonlinear. Figure 5–11B shows a shunt ohmmeter scale, which is explained next.

In the series ohmmeter, the unknown resistance is connected in series with the meter movement. In the shunt ohmmeter, the unknown resistance is connected in shunt, or in parallel, with the meter movement, as shown in Figure 5–12. Once again, there is a range resistor R_1, a zero resistor R_z, a meter resistance R_m, and a battery V_T. There is also an on/off switch since a closed circuit results without the unknown resistance connected.

Another difference between the two ohmmeters is that the shunt ohmmeter has zero on the left and infinity (∞) on the right end of the scale, as shown in Figure 5–11B. When the meter is turned on, current flows and the zero control is adjusted for full-scale deflec-

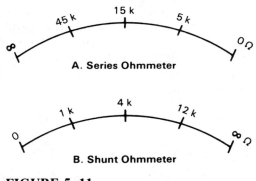

A. Series Ohmmeter

B. Shunt Ohmmeter

──O **FIGURE 5–11**
Ohmmeter Scales

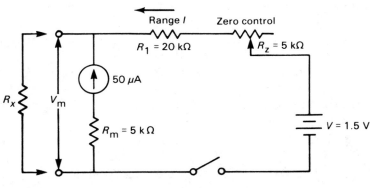

──O **FIGURE 5–12**
Shunt Ohmmeter Circuit

tion. Shorting the input causes the current to bypass the movement and results in no pointer deflection. Resistance values between short and open cause deflections between zero and maximum.

○ Example of Calculations for the Shunt Ohmmeter

The amount of resistance for specific amounts of deflection in the shunt ohmmeter is calculated in the following manner. With no resistor applied ($R_x = \infty\ \Omega$), the total current is as follows:

$$I_T = \frac{V_T}{R_T} = \frac{V_T}{R_1 + R_m + R_z} = \frac{1.5\ \text{V}}{20,000\ \Omega + 5000\ \Omega + 5000\ \Omega}$$

$$= 50\ \mu\text{A}$$

For 1/2-scale deflection, 25 microamperes are required. The calculations are as follows:

$$V_m = I_m \times R_m = 0.000025 \text{ A} \times 5000 \text{ } \Omega = 0.125 \text{ V}$$

$$V_1 + V_z = V_T - V_m = 1.5 \text{ V} - 0.125 \text{ V} = 1.375 \text{ V}$$

$$I_T = \frac{V_1 + V_2}{R_1 + R_2} = \frac{1.375 \text{ V}}{20,000 \text{ } \Omega + 5000 \text{ } \Omega} = 0.000055 \text{ A}$$

$$I_x = I_T - I_m = 0.000055 \text{ A} - 0.000025 \text{ A} = 0.000030 \text{ A}$$

$$R_x = \frac{V_m}{I_x} = \frac{0.125 \text{ V}}{0.000030 \text{ A}} = 4166 \text{ } \Omega$$

For 1/4-scale deflection, 12.5 microamperes are required. The calculations are as follows:

$$V_m = I_m \times R_m = 0.0000125 \text{ A} \times 5000 \text{ } \Omega = 0.0625 \text{ V}$$

$$V_1 + V_z = V_T - V_m = 1.5 \text{ V} - 0.0625 \text{ V} = 1.4375 \text{ V}$$

$$I_T = \frac{V_1 + V_2}{R_1 + R_2} = \frac{1.4375 \text{ V}}{25,000 \text{ } \Omega} = 0.0000575 \text{ A}$$

$$I_x = I_T - I_m = 0.0000575 \text{ A} - 0.0000125 \text{ A} = 0.000045 \text{ A}$$

$$R_x = \frac{V_m}{I_x} = \frac{0.0625 \text{ V}}{0.000045 \text{ A}} = 1389 \text{ } \Omega$$

For 3/4-scale deflection, 37.5 microamperes are required. The calculations follow:

$$V_m = I_m \times R_m = 0.0000375 \text{ A} \times 5000 \text{ } \Omega = 0.1875 \text{ V}$$

$$V_1 + V_z = V_T - V_m = 1.5 \text{ V} - 0.1875 \text{ V} = 1.3125 \text{ V}$$

$$I_T = \frac{V_1 + V_2}{R_1 + R_2} = \frac{1.3125 \text{ V}}{25,000 \text{ } \Omega} = 0.0000525 \text{ A}$$

$$I_x = I_T - I_m = 0.0000525 \text{ A} - 0.0000375 \text{ A} = 0.000015 \text{ A}$$

$$R_x = \frac{V_m}{I_x} = \frac{0.1875 \text{ V}}{0.000015 \text{ A}} = 12,500 \text{ } \Omega$$

Multirange ohmmeters use more than one battery. Typically, 1.5 volts are used for most ranges. On the highest-resistance range, 15 or 30 volts are used in order to produce a measurable deflection.

ANALOG MULTIMETER SPECIFICATIONS

Most VOMs are battery-operated and have phenolic cases with handles, making them both portable and durable. The differences between the many brands and models depend on four basic specifications: accuracy, sensitivity, frequency response, and the number of ranges. The first three specifications will be discussed here; the last is self-explanatory.

Accuracy

The *accuracy* specification of a meter describes how close the meter will indicate the true or actual value. Voltmeter and ammeter accuracies are expressed as a percentage of the full-scale value. For example, a 3% voltmeter will indicate any value on the 100-volt range within 3% of 100 volts, or within ±3 volts. Actually, a better name for this specification would be *error,* since error is what is being described.

While accuracy is usually expressed as a percentage of the full-scale value, there is another consideration. Suppose you were measuring 20 volts on the 100-volt range of a 3% meter. You can expect an accuracy of 3% of the full-scale range of 100, or an indication within 3 volts. In other words, you can measure 20 volts within ±3 volts. Now, consider the actual or true accuracy.

$$\text{true accuracy of indication} = \frac{3 \text{ V}}{20 \text{ V}} \times 100\% = 15\%$$

In reality, the indication is within 15%, not 3%. However, if you use the 50-volt range, the results will improve.

$$3\% \text{ of } 50 \text{ V} = \pm 1.5 \text{ V}$$

$$\text{true accuracy of indication} = \frac{1.5 \text{ V}}{20 \text{ V}} \times 100\% = 7.5\%$$

The point to remember is, *use the lowest range possible for the most accurate voltage and current measurements.*

Ohmmeter accuracy is usually expressed in degrees or as a percentage of full-scale arc. Figure 5–13 shows the significance of a ±2% error of full-scale arc in a 35-ohm resistance indication. So,

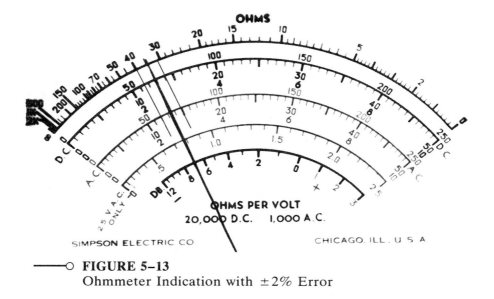

──○ **FIGURE 5–13**
Ohmmeter Indication with ±2% Error

*use the range that places the pointer nearest the center for the most
accurate resistance measurements.*

Typical full-scale accuracy specifications vary from 1% to 3%
for DC voltage and current. Accuracy for AC voltage is typically
±2% to 4%. Resistance accuracy is generally ±2% to 4%.

Sensitivity

The DC voltage *sensitivity* of a VOM is expressed in ohms per volt,
and it is the reciprocal of the full-scale deflection current. It describes how an instrument appears to a circuit under test. Consider
the meter circuit in Figure 5–4. The meter is a 1-volt FSD meter,
and the circuit has a total resistance of 20,000 ohms. So, this meter
has a sensitivity of 20,000 ohms per volt.

Now, consider the multirange voltmeter circuit in Figure 5–5A.
When the meter is on the 5-volt range and connected across a circuit,
that circuit has 100,000 ohms placed in parallel with it. The sensitivity of this meter is as follows:

$$\frac{100,000\ \Omega}{5\ V} = 20,000\ \Omega/V$$

This meter has a sensitivity of 20,000 ohms per volt for all ranges.

While sensitivity is a constant for all ranges, the apparent resistance of a meter, as seen by a circuit under test, changes with each range. The significance of sensitivity is illustrated in the following example.

○ Example of Sensitivity

As shown in Figure 5–14A, 5 volts are dropped across each resistor in a circuit. When the meter is switched to the 5-volt range and connected across R_2, as in Figure 5–14B, it appears as 100 kilohms connected across the 100-kilohm resistor R_2. This connection produces a series-parallel circuit of 50 kilohms in series with the 100-kilohm resistor R_1.

To see how a meter can affect a circuit, we begin with the calculation of expected values. Consider the circuit in Figure 5–14A. The expected voltage V_2 is as follows:

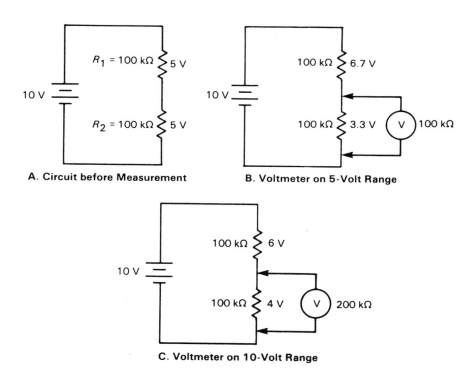

A. Circuit before Measurement   B. Voltmeter on 5-Volt Range

C. Voltmeter on 10-Volt Range

──○ **FIGURE 5–14**
Circuit under Test

$$V_2 = \frac{R_2 \times V_T}{R_1 + R_2} = \frac{100 \text{ k}\Omega \times 10 \text{ V}}{200 \text{ k}\Omega} = 5 \text{ V}$$

If we switch the meter to its 5-volt range and connect it across R_2, our circuit will resemble the circuit of Figure 5–14B. The voltage across R_2 now can be recalculated as follows:

$$V_2 = \frac{50 \text{ k}\Omega \times 10 \text{ V}}{150 \text{ k}\Omega} = 3.3 \text{ V}$$

As we can see, the meter has introduced considerable error because it has changed the parameters of the circuit. The voltage across R_2 has been reduced from 5 volts to 3.3 volts because of the additional parallel circuit caused by the meter. This effect is called *meter loading,* and it varies with meter sensitivity and range.

Now, consider what happens when the 10-volt range is used, as shown in Figure 5–14C. This 20,000-ohm-per-volt meter has a resistance of 200,000 ohms when on the 10-volt range. Connecting this 200 kilohms across the 100 kilohms (R_2) produces an equivalent resistance of 66,667 ohms. So, V_2 is as follows:

$$V_2 = \frac{66,667 \ \Omega \times 10 \text{ V}}{166,667 \ \Omega} = 4 \text{ V}$$

The error is less, although it still is significant.

A point to remember is, *the effects of meter loading decrease as the meter range is increased.* However, this point contradicts the rule about accuracy. There are various ways to resolve this dilemma.

Be aware of the characteristics of the circuit you are measuring, and determine whether loading is likely to be a concern. If it is, use a high-sensitivity meter. Typical values for sensitivity range from 5000 to 20,000 ohms per volt. High-impedance circuits are susceptible to loading and quite often cannot be measured with some VOMs. Electronic VOMs, described later in this chapter, are needed in these situations.

Frequency Response

The frequency range over which an AC voltmeter can meet its accuracy specifications is called its *frequency response.* Frequency response is another factor you must consider in the selection of any measurement instrument.

At very low frequencies, a meter pointer tends to oscillate as it attempts to follow the pulses. As the frequency increases, meter response varies according to range. A typical analog VOM can accurately measure voltages at frequencies up to 1 kilohertz on high-voltage ranges and up to 100 kilohertz on low-voltage ranges. Above the specified limit, the pointer indication begins to drop.

Waveform Error

Another problem in the measurement of AC voltages results from the waveshape of the applied signal. Many VOMs operate with circuits and calibrations that are accurate only with pure sine waves. However, signals are often not pure sine waves. Recall that a PMMC meter responds to the average of the current through its coil. Since AC alternates around zero, the average value of a sine wave is zero. So, in order to measure a sine wave, we must change it to DC. In fact, a sine wave is generally changed to pulsating DC with meter rectifiers. Then, the pointer responds to the average value but indicates the rms value.

Consider an AC voltage of 2 volts peak-to-peak (pp). Full-wave rectifiers change it to a pulsating signal with a peak value of 1 volt. Then, the PMMC system averages it as 0.637 times peak, or 0.637 volt. However, since a sine wave has an rms value of 0.707 times peak, or 0.707 volt, the scale is calibrated accordingly. So, a 2-volt-pp voltage will produce an indication of 0.707 on the scale of the meter because of the process normally used. Once again, the meter rectifies, averages, and multiplies the average by 0.707/0.637, or about 1.11.

While the 0.707/0.637 relationship is true for sinusoidal wave-forms—and they are the most common waveforms—it is not true for all waveshapes. Consider what occurs when a square wave is measured. The average value of a full-wave, rectified, 2-volt-pp square wave is 1 volt. Its rms value is also 1 volt. However, the typical PMMC meter would rectify, average, and then multiply it by 1.11. The meter would incorrectly indicate 1.11 volts rms instead of the correct value of 1.0 volt rms.

This 11% error is caused by the incorrect assumption about the waveform, an assumption that is valid only for sine waves. The assumption is that an rms value is equal to 1.11 times its average value. For this reason, the problem is called *waveform error*. The amount of error depends on the waveshape of the signal, and it can exceed

55% for pulsed voltages. To avoid waveform error, use an instrument that measures true rms values.

ANALOG MULTIMETER APPLICATIONS

Measuring a parameter with a VOM begins with the decision that the instrument is appropriate for the task. Considerations in this decision include the magnitude and the frequency of the parameter and the required accuracy. When you are sure that an instrument is adequate, the measurement process can begin.

Voltage Measurement

While measuring a voltage is a simple process, a safe routine should always be followed. The following steps are suggested:

1. Connect the leads to the instrument.
2. Place the mode switch in the appropriate position (volts and DC or AC).
3. Select the highest voltage range.
4. Connect the ground or common lead to the circuit being measured.
5. Connect the positive lead to the circuit.
6. Energize the circuit (if it is not already energized).
7. Switch down the ranges, one range at a time, until you reach the lowest usable range (with the pointer still on the scale). This step should be done by removing the positive lead, switching down one range, and replacing the lead.
8. Read the indicated value.
9. Disconnect the leads from the circuit and switch to off or to a high-voltage range. Leaving the meter switched to a high-voltage range reduces the vulnerability of the meter.

During this process, you should consider sensitivity and the possibility of circuit loading. Remember, also, that most errors in voltage measurement are caused by misinterpretation of the indicated value. For example, the graduations used depend on the range selected. The value indicated in Figure 5–15 is 65 volts on the 250-

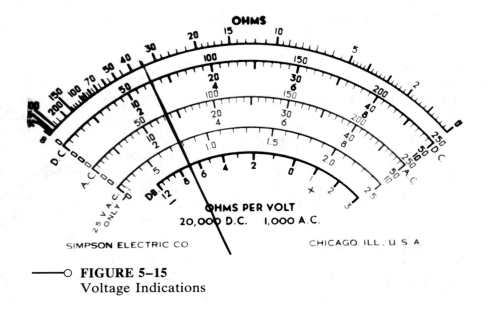

————○ **FIGURE 5–15**
Voltage Indications

volt-DC range, 67 volts on the 250-volt-AC range, and 0.79 on the 2.5-volt-AC range. These differences occur because the three scales have different linearities.

Current Measurement

The process of current measurement is similar to the process of voltage measurement, with one significant difference. *The circuit must be broken for current measurement.* While this statement is not true for a clamp-on type of ammeter, it is for all others, which happen to be the great majority of ammeters in use. So, remember that you *measure voltage across and current through a circuit.* Also, a circuit should be de-energized before you connect the meter, energized for the measurement, and de-energized before you disconnect the meter.

The following procedure is recommended for current measurement with a VOM:

1. Connect the leads to the instrument.
2. Select DC amperes or milliamperes and the highest range.
3. De-energize the circuit to be measured.

4. Break the circuit at the point the current is to be measured.
5. Connect the instrument probes to the separated wires with minus (−) toward the negative part of the circuit and plus (+) toward the positive part, as shown in Figure 5–16.
6. Energize the circuit.
7. Switch down the ranges to the point of highest value on the scale and pointer deflection. De-energize the circuit or disconnect the meter while you are switching ranges.
8. Read the indicated value and de-energize the circuit.
9. Disconnect the probes and reconnect the circuit.

Resistance Measurement

Ideally, a component should be out of a circuit when its resistance is measured. When separation is not possible, the circuit must be de-energized, and one lead of the component should be disconnected from the remainder of the circuit. Otherwise, parallel current paths or alternate current sources can be involved, causing incorrect results and possible instrument damage.

Once the power is off and one lead is disconnected, take the following steps:

1. Select the ohms function and the $R \times 100$ range.
2. Zero the meter by touching the probe tips together and adjusting the zero adjust control for an indication of 0 ohms.
3. Connect the probes to the component being measured. Do not hold their metal tips, since doing so would introduce you as a parallel path.

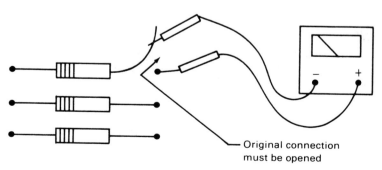

Original connection must be opened

FIGURE 5–16
Ammeter Connection for Current Measurement

4. Select the range with the resistance indication closest to midscale.
5. If the range selected in step 4 is not the $R \times 100$ range (which you have already zeroed), disconnect the probes and zero the meter for this range.
6. Reconnect the probes to the component and read the indicated value.
7. Disconnect the probes and reconnect the circuit.
8. Switch to off or to a high-voltage range.

To check diodes or transistors, you place a VOM on a resistance range and measure the junction resistance in both directions. It should be high in one direction and low in the other. For a signal diode, less than 100 ohms in the forward direction and more than 100,000 ohms in the reverse direction are reasonable limits—that is, a front-to-back ratio of 1000 to 1. Power diodes will probably show lower values. Remember, this procedure is a *check* and not a comprehensive measurement. Your purpose is simply to find a diode that may be shorted or opened.

A diode check involves two other concerns: meter voltage and current. The meter test voltage could be below the forward voltage of the diode, causing it to appear open. It could also exceed the reverse breakdown voltage and damage the junction. The range selected could cause excessive current to flow in the forward direction. Once again, read the manual of the meter and follow the manufacturer's recommendations.

Decibel Measurement

The *decibel* is a unit used to reference one voltage or power level to another. While various specific units exist, the common unit is the decibel-milliwatt (dBm). It is measured as a specified voltage output into a specified load. An output of 1 decibel-milliwatt equals 0.001 watt into a load of 600 ohms. Most VOMs have a decibel scale. The decibel scale in Figure 5–15 is located on the bottom and ranges from −12 to +3 decibels.

While the procedure for decibel measurement may vary from one instrument to another, the following steps are typical. Connect the probes to the output terminals of the properly loaded circuit, as if you were measuring AC voltage. Select the best range by using the output function. If the best range is 2.5 volts, you can read the decibels directly from the decibel scale. If, however, the best range is higher, you must add a decibel amount, as specified in the operator's manual.

The procedures outlined in this section generally apply to most instruments. An understanding of these procedures is necessary in your work as a technician. In addition, you have one other important responsibility: Read the instruction manual before operating any equipment.

OTHER ANALOG INSTRUMENTS

The preceding sections described the most common analog instruments. Four additional analog meters are described here. Some can measure voltage and current without the problems of loading or waveform error. One meter is used for the direct measurement of power.

Electronic VOM

An electronic multimeter, like the one in Figure 5–17, has an amplifier circuit added between the circuit under test and the typical meter movement. This input-stage amplifier means that the circuit under test no longer supplies the power for pointer displacement. Instead, the input signal is used as a reference, and the amplifier drives the pointer to a position proportional to the input signal. Closed-loop and differential amplifier circuits are commonly used in this application. Because of their high input impedance, field effect transistors (FETs) are generally used.

The basic electronic voltmeter circuit in Figure 5–18 provides a constant 1-megohm load to an incoming signal. The range switch divides that signal down to a level of 1 volt maximum. This signal then proceeds to a noninverting operational amplifier with a gain of 1. The amplifier, in turn, drives the meter movement. Diodes are often placed between the amplifier input and the ground as signal amplitude limiters. Diodes are also used to rectify AC voltages at the DC meter movement. More will be said about electronic multimeters in the next chapter.

Three features should be noted about electronic multimeters, often called FET VOMs:

1. The electronic front end increases the input impedance to a constant value for all ranges, generally 1 or 10 megohms.

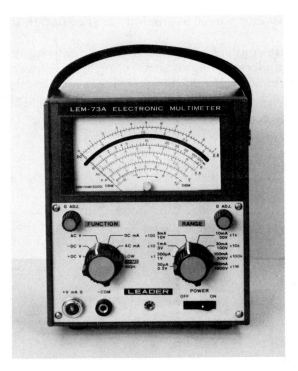

———○ **FIGURE 5–17**
Electronic Multimeter (Photo courtesy of Leader Instruments Corporation)

2. The upper frequency limit is often extended to 3 megahertz or above.
3. The voltage placed across components during a resistance measurement can be lower than the voltage from a nonelectronic VOM. When you are checking the resistance of semiconductors, the voltage applied by the meter may not be enough to exceed the threshold voltage.

Thermocouple Meter

Waveform error is a problem with some instruments. They average nonsinusoidal voltages when we actually want to know the rms or effective value. One solution to this problem is to use a meter designed around a thermocouple, as indicated in Figure 5–19. It works as follows: Current to be measured is directed through a conductive

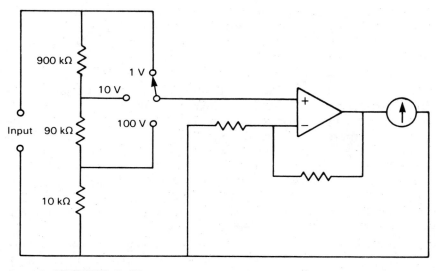

—○ **FIGURE 5–18**
Electronic Voltmeter Circuit

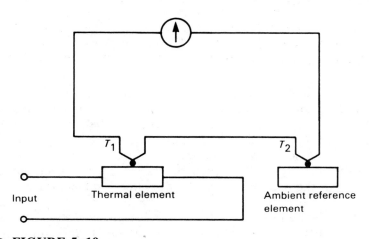

—○ **FIGURE 5–19**
Thermocouple Transfer Instrument

metal block, causing the block's temperature to rise in proportion to the square of the current (I^2R).

The thermocouple meter responds to the *effective* value of any current, since effective value is what causes the temperature to rise. The effective value of direct current is the average, while the effective value of alternating current is its rms value. Remember, DC average has the same heating effect as AC rms. Since this meter can measure DC or AC directly without rectification, it is called a *transfer instrument*.

In Figure 5–19, thermocouple 1 (T_1) is thermally, but not electrically, attached to the block or heating element. It consists of dissimilar metals bonded together at the point of contact with the block. This thermocouple produces a small voltage proportional to the junction temperature. Thermocouple 2 (T_2) develops a voltage proportional to the ambient temperature. The difference between these voltages represents the heating caused by the current. This difference voltage is applied to a sensitive meter with a hand-drawn current scale.

Accuracy for the thermocouple meter exceeds 1%, and the frequency range is greater than 50 megahertz. While they are basically current-measuring instruments, thermocouple meters can measure voltage if precision multipliers are added to their circuits. Thermocouple instruments are generally used as working standards. Care must be taken when you are working with them because they respond slowly, are intolerant of overload, can be damaged easily, and are costly to repair.

Electrodynamometer

Another type of transfer instrument is the *electrodynamometer*. It uses an electromagnet in place of the permanent magnet. Recall that a PMMC meter operates on the interaction between the magnetic field produced by coil current and the field of the permanent magnet. A problem arises when we measure AC with the PMMC meter because the electromagnetic field reverses polarity as the current alternates but the permanent magnet does not. An electrodynamometer resolves that problem.

As indicated in Figure 5–20, the current being measured flows through two fixed coils and one moving coil. Pointer movement is unlike the movement of a PMMC meter in two ways. First, all fields reverse when the current reverses. Thus, the repulsion will remain and not become attraction every half cycle. Second, since both fixed and moving fields are caused by the applied current, the deflection is proportional to the square of the current.

Since current must travel through three coils rather than one, this meter has limitations. Inductive reactance limits frequency response to the low–audio frequency range. Power consumption is

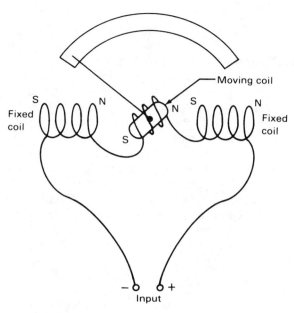

○ **FIGURE 5–20**
Electrodynamometer Movement

higher than in PMMC meters, causing a sensitivity below 100 ohms per volt. However, the multicoil system has advantages beyond the measurement of true rms. Connecting the fixed coils in series with current and the moving coil (with a multiplier) in parallel with voltage allows this meter to respond to circuit power and become a wattmeter.

Power factor can be measured with an electrodynamometer having two fixed (current) coils and two moving (voltage) coils. The moving coils are mounted on the same shaft but positioned 90 degrees apart. Currents are phased 90 degrees apart by adding a series resistor to one moving coil and a series inductor to the other.

The power factor of a circuit is based on the phase relationship between circuit voltage and circuit current. With a power factor of 100%, there is no phase shift between circuit current and voltage. The coil reacts primarily with the in-phase (resistive) voltage coil since the fields of these two coils are in phase with each other. However, with a low power factor, current lags voltage, causing a lagging field around the current coil. In this case, the primary magnetic interaction is with the inductive voltage coil, producing a different pointer position.

SUMMARY

One of the most common analog meters is the VOM with taut-band suspension. This system uses a moving coil supported by a twisted, stretched ribbon that provides support with little friction. Pointer displacement is proportional to average current. Voltage is measured with series multiplier resistors, and current is measured with parallel shunt resistors. Resistance is measured by using a circuit of batteries and a voltage divider.

Common specifications of an analog VOM are accuracy, sensitivity, frequency response, and number of ranges. Typical DC voltage and current accuracy is 1% to 3%, while AC voltage accuracy is 2% to 4%. Resistance accuracy is typically specified as within 2% to 4%. Since voltage and current accuracy are described as percentages of full scale, true accuracy must be calculated. This calculation is done by dividing the specified accuracy by the value measured. True accuracy can exceed 40%. The lowest true accuracy is achieved by using the lowest possible range.

Sensitivity is expressed in ohms per volt and describes the degree of possible loading. It is a function of the coil and multiplier resistances. Typical values are 5000 to 20,000 ohms per volt. Frequency response describes the range over which an AC voltage can be accurately measured. A typical upper limit for high-voltage is 1 kilohertz, and a typical limit is 100 kilohertz for lower ranges.

Voltage is measured with a VOM by connecting the instrument across the voltage source. The lowest safe range provides the most accuracy. The greatest cause of voltage measurement error is caused by misreading the scale.

Current is measured by inserting the meter in series with the circuit. Once again, the best range is the lowest safe range.

Resistance measurement requires the removal of power and the disconnection of one of the component's leads. The ohmmeter must be zeroed on the range used. The best range to use for resistance is the one that places the pointer nearest the center of the scale. Decibel measurements are made by switching the mode to AC and connecting the meter across the circuit.

Electronic multimeters add amplifier circuits between input signal and meter movement. These amplifiers provide the power for pointer deflection, reducing the demands on the circuit under test and producing an input impedance in excess of 10 megohms. Thermocouple meters directly measure the effective values of DC or AC by responding to their heating effect. Since they can directly measure

AC and DC, thermocouple meters are called transfer instruments and are often used as working standards.

Electrodynamometers use an electromagnet instead of a permanent magnet. All fields reverse when the applied current reverses, so no rectifiers are needed in order to measure true rms. Although this instrument has low sensitivity and low frequency response, it has its advantages. With changes in coil connection or positioning, this instrument can measure circuit power or power factor.

BIBLIOGRAPHY

Basic Electrical Measurements, Melville B. Stout, Prentice-Hall, 1950.

Electronic Instrumentation and Measurement Techniques, 2nd ed., William David Cooper, Prentice-Hall, 1978.

Handbook of Electronic Meters, Theory and Application, John D. Lenk, Prentice-Hall, 1981.

Fundamentals of Electronics, Douglas R. Malcolm, Jr., Breton Publishers, 1983.

QUESTIONS

1. Define *accuracy, sensitivity,* and *frequency response.*
2. Define *transfer instrument,* and give an example.
3. Define *waveform error,* and describe how it can be eliminated.
4. Compare the characteristics and specifications of electronic and nonelectronic VOMs.
5. Compare taut-band and jeweled-bearing suspensions.
6. Describe the operation of a thermocouple ammeter.
7. Describe the operation of an electrodynamometer.
8. Name and describe the four basic specifications of an analog VOM.
9. Explain why a make-before-break switch is used with some multirange ammeters.
10. Describe the primary advantage of an Ayrton shunt over the basic multirange ammeter.

11. What is the common indicator that an ohmmeter has a shunt or a series circuit?

12. Explain the operational difference between a series and a shunt ohmmeter.

13. Explain how a PMMC meter measures AC voltage and current.

14. What is the primary factor that limits the frequency response of a PMMC voltmeter?

15. Why do most electric VOMs have a separate scale for 2.5 volts AC and below?

16. Why is a zero control needed in ohmmeters?

17. Describe ohmmeter accuracy.

18. Which is the best range to use when true accuracy is the primary concern?

19. Which is the best range to use on a voltmeter when loading is a concern?

PROBLEMS

1. Determine the multiplier values for the 1-, 50-, and 150-volt ranges of a voltmeter with a 200-microampere, 5000-ohm movement.

2. Determine the multiplier values for the 5-, 20-, and 100-volt ranges of a voltmeter with a 200-microampere, 5000-ohm movement.

3. Determine the multiplier values for the voltmeter in Figure 5–5A so that it has 2.5-, 25-, and 50-volt ranges.

4. Determine the multiplier values for the voltmeter in Figure 5–5B so that it has 0.2-, 1-, 5-, and 10-volt ranges with a 2000-ohm, 100-microampere movement.

5. Determine the shunt values for the 500-microampere, 2-milliampere, and 10-milliampere ranges of an ammeter with a 100-microampere, 2000-ohm movement.

6. Determine the shunt values for the 500-microampere, 1-milliampere, and 5-milliampere ranges of an ammeter with a 200-microampere, 5000-ohm movement.

7. Design an Ayrton shunt ammeter with a full-scale deflection of 0.2, 0.5, and 1.0 milliampere for a 5000-ohm, 100-microampere movement.

8. Design an Ayrton shunt ammeter with a full-scale deflection of 0.1, 0.5, and 1.0 milliampere for a 2000-ohm, 50-microampere movement.

9. Determine the values of R_1 and R_z for a series ohmmeter with a 5000-ohm, 100-microampere movement and a 1.5-volt battery.

10. Determine the values of R_1 and R_z for a series ohmmeter with a 2000-ohm, 50-microampere movement and a 1.5-volt battery.

11. Determine the values of R_1 and R_z for a shunt ohmmeter with a 5000-ohm, 100-microampere movement and a 1.5-volt battery.

12. Determine the values of R_1 and R_z for a shunt ohmmeter with a 2000-ohm, 50-microampere movement and 1.5-volt battery.

13. Calculate the 1/4-, 1/2-, and 3/4-scale deflection resistances for the ohmmeter described in Problem 9.

14. Calculate the 1/4-, 1/2-, and 3/4-scale deflection resistances for the ohmmeter described in Problem 10.

15. Calculate the 1/4-, 1/2-, and 3/4-scale deflection resistances for the ohmmeter described in Problem 11.

16. Calculate the 1/4-, 1/2-, and 3/4-scale deflection resistances for the ohmmeter described in Problem 12.

17. Calculate the true accuracy of a 32-volt indication on the 100-volt range of a 3% FSD meter. Repeat the calculation for the 50-volt range.

18. Calculate the true accuracy of a 14-volt indication on the 50-volt range of a 2% FSD meter. Repeat the calculation for the 25-volt range.

19. Calculate power and the worst-case error by using the following measurements, which were taken with a 3% FSD meter: V = 18 volts on the 50-volt range and I = 250 milliamperes on the 500-milliampere range.

20. Calculate power and the worst-case error by using the following measurements, which were taken with a 2% FSD meter: V = 6 volts on the 10-volt range and I = 40 milliamperes on the 50-milliampere range.

21. Determine the loading error when you measure the voltage across a 200-kilohm resistor connected in series with another 200-kilohm resistor and a 15-volt source. The meter has a sensitivity of 20,000 ohms per volt and is on the 10-volt range.

22. Determine the loading error when you measure the voltage across a 1-megohm resistor connected in series with a 500-kilohm resistor and a 30-volt source. The meter has a sensitivity of 10,000 ohms per volt and is on the 25-volt range.

6

Digital Meters

OBJECTIVES

Digital meters are quickly becoming the standard of the industry. Through microelectronics, they have more complex circuitry with fewer discrete components than their analog counterparts. They also offer greater accuracy, more durability, and additional capabilities such as the ability to show true rms values. This chapter describes their theory of operation, specifications, and applications.

Name and describe three functional sections of a digital voltmeter.

Describe the signal preparation in a digital multimeter.

Explain the differences between *average* and *effective* (true rms) circuits in measurement instruments.

Name and describe five common specifications of a digital multimeter.

Measure voltage, current, and resistance with a digital multimeter.

DIGITAL VOLTMETER THEORY

A typical digital multimeter is pictured in Figure 6–1. Digital multimeter operation is designed around the digital voltmeter circuit. That is, a parameter is represented as a voltage and then measured.

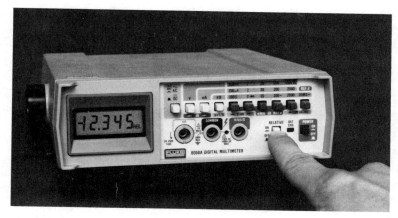

—○ **FIGURE 6–1**
Digital Multimeter (Photo courtesy of John Fluke Mfg. Co., Inc.)

The steps in the measurement process, illustrated by the block diagram in Figure 6–2, involve protecting the source from loading, increasing or decreasing the signal to a measurable size, converting it to DC if necessary, changing it from analog to digital form, and displaying its magnitude with light-emitting diodes (LEDs) or liquid crystal displays (LCDs). Our discussion begins with the first few steps in that list.

Signal Preparation

The two concerns at the signal preparation stage are signal amplitude and the avoidance of source loading. An incoming signal that is too large can be reduced with an *attenuator*—a passive circuit usually consisting of resistors—as in a basic analog VOM. A signal that is too small can be increased with an *amplifier*—a circuit in which a small current or voltage controls a large current or voltage—as in an electronic VOM. Both processes are used in digital voltmeters.

—○ **FIGURE 6–2**
Block Diagram for Digital Voltmeter

One technique for signal preparation is a *voltage divider* circuit like the circuit in Figure 6–3A, which offers a constant 10-megohm load to the signal source. The voltage divider is simply a series resistive circuit across which the total input voltage is applied (volts to common). Then, reduced portions of the total voltage can be taken from the divider and sent to the circuitry that follows it. This divider appears as a constant load to the signal source, even when

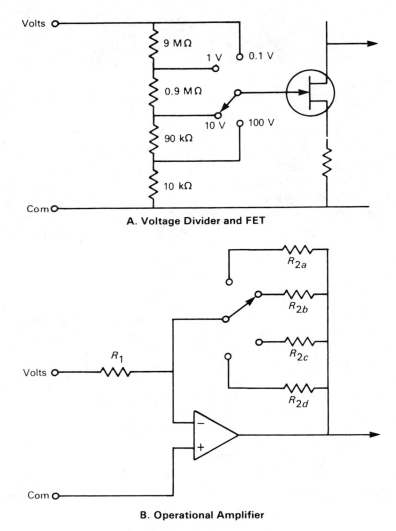

A. Voltage Divider and FET

B. Operational Amplifier

FIGURE 6–3
Input Circuits for Digital Voltmeter

the meter is on the lowest voltage range and the FET is directly across the divider, because of the high impedance of the FET gate. The reduced input voltage, when applied to the FET gate, controls the FET channel current and, in turn, drains voltage. This voltage then proceeds on to the ADC. Like other electronic meters, this meter isolates the signal source from the indicator drivers in order to avoid loading and changing the parameter being measured. Incoming signals that are too large are reduced with the divider, and signals that are too small are amplified with the FET.

The circuit in Figure 6–3B utilizes an operational amplifier and feedback to accomplish what the divider and the FET can do. Typically, an *operational amplifier* is a multistage, integrated circuit with high-impedance differential inputs and a low-impedance output. It is direct-coupled and offers high gain and wideband operation. Amplification is controlled with negative feedback, since the *gain* of the amplifier is proportional to the ratio of the input and feedback resistors. That is:

$$\text{amplifier gain} = A = \frac{R_2}{R_1}$$

Both of the circuits in Figure 6–3 allow signals from microvolts to kilovolts to be scaled to a measurable size with accuracy and stability. In many digital voltmeters, the circuits do the scaling automatically. This feature is called *autoranging*.

Figure 6–4 shows one method by which autoranging can be accomplished. When a small signal is applied, both FETs are off, producing high resistances in the R_2 branches. With a high R_2/R_1 ratio, the amplifier provides a high gain and makes the incoming signal large. If, however, the incoming signal is already large, the amplifier output could become excessive. This excessive output would be detected by overflow circuitry.

The overflow circuit feeds back signals to the appropriate FET, and it begins to conduct. Operation in the FET reduces the effective R_2 resistance in the R_2/R_1 equation to the amount in that particular branch. In turn, the gain of the amplifier is reduced accordingly, and the large incoming signal no longer overdrives the display limits.

Analog-to-Digital Conversion

The next step in the meter's measurement process is to take the scaled analog voltage and convert it to digital data. Typically, it will be converted to binary or binary-coded decimal (BCD) data through the use of a single chip with large-scale integration (LSI).

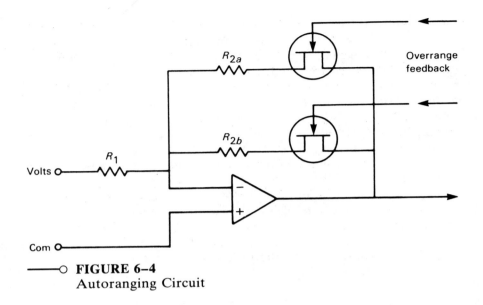

——○ **FIGURE 6–4**
Autoranging Circuit

Although digital voltmeter design is undergoing rapid technological innovation, three basic analog-to-digital converter (ADC)
systems are common. These systems are the ramp ADC, the dual-
slope ADC, and the successive-approximation ADC.

The simple *ramp ADC* is shown in Figure 6–5. A constant-
current source provides power to a high-quality *RC* circuit, which
serves as a ramp generator. With this source applied, the capacitor
begins to charge. The linear portion of the capacitor voltage ramp
is passed on to the two comparators. One comparator also looks at
the voltage to be measured, while the other also looks at ground.

Elsewhere in the system is a crystal-controlled *oscillator-divider
circuit,* which produces a specific number of discrete pulses (for example, 1000 per second). The counting time is controlled by the gate,
and the count is passed on to a display system.

Figure 6–6 shows the sequence of events in the voltage comparison. The *comparators* of Figure 6–5 are used to sense specific
voltage levels. First, the ground comparator recognizes when the
ramp, biased below 0 volts, crosses 0 on the way up (point 1). It
then informs the gate to start the counter. Second, the input comparator recognizes when the ramp voltage equals the input voltage
(point 2). This comparator informs the gate to stop the counter. The
displayed count represents the unknown voltage.

This system works because of the way in which it is calibrated.
In calibration, an accurate 1-volt signal is entered directly into the
comparator. It is then compared with the ramp, and the *RC* time

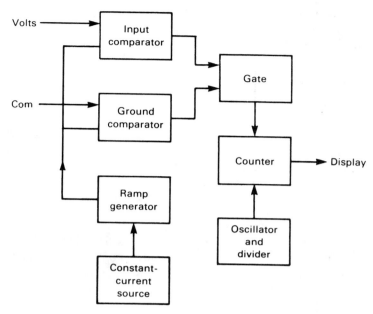

—○ **FIGURE 6–5**
Ramp ADC

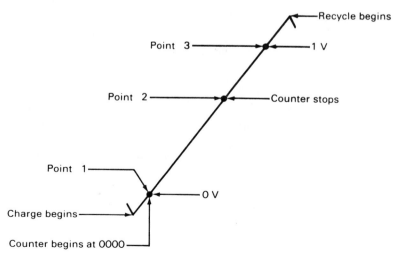

—○ **FIGURE 6–6**
Points of Voltage Comparison

constant is adjusted so that the ramp equals 1 volt when the counter reaches 1000. Since the ramp is linear, 0.5 volt would occur when a count of 500 occurred, and so on. Because of the linearity of the curve, the stability of the oscillator, and the accuracy of the voltage standard, this system is easily accurate to 0.1%.

Another common ADC system is the *dual-slope ADC,* or dual-slope integrator, shown in Figure 6–7. Where charge time was used in the ramp ADC, discharge time is used here. In this system, the unknown voltage is used to charge a capacitor for a controlled period of time. Then, the unknown voltage is separated from the ramp circuit, and a constant-current source is connected. The ramp generator capacitor is then discharged through the current source. The beginning and the end of the discharge cycle is used, through the gate, to start and stop the counter.

The dual-slope ADC is used more often than the ramp ADC, and it has certain advantages. In the dual-slope ADC, changes in R or C (called transients) that affect both charging and discharging time and that produce errors are canceled. Noise is also canceled in the dual-slope ADC. However, there is a significant limitation to both ADCs. Both are slow in operation and are limited to few samples per second.

The *successive-approximation ADC* shown in Figure 6–8 is faster than the ramp or the dual-slope ADC. In this system, a sample of the unknown voltage is taken and held in a register (sample holder) for a short period of time. Accurate levels of decreasing voltages are

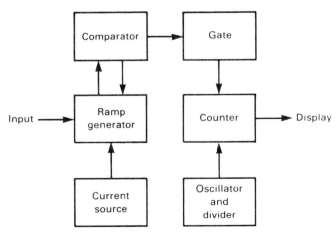

—○ **FIGURE 6–7**
Dual-Slope ADC

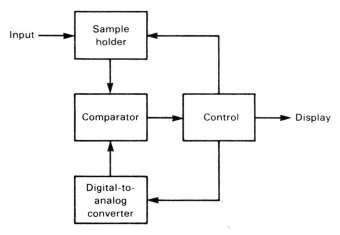

──○ **FIGURE 6–8**
Successive-Approximation ADC

briefly sent from the digital-to-analog converter (DAC) to the comparator and compared with the sample of the unknown voltage.

For this discussion, let us consider the unknown voltage to have a value of 10.5 volts. A series of comparisons, shown in Figure 6–9, proceed as follows. The control (of Figure 6–8) instructs the

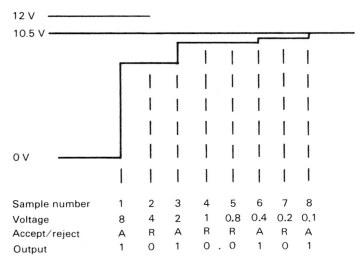

Sample number	1	2	3	4	5	6	7	8
Voltage	8	4	2	1	0.8	0.4	0.2	0.1
Accept/reject	A	R	A	R	R	A	R	A
Output	1	0	1	0	. 0	1	0	1

──○ **FIGURE 6–9**
Points of Successive Approximation

DAC to send test pulse 1, which is 8 volts. This pulse is less than the unknown voltage, so it is accepted. The comparator sends a signal for the number 1 to the control. Pulse 2 is sent next, and it is 4 volts. Since 8 + 4 exceeds the unknown voltage, this pulse is rejected, and the signal for the number 0 is sent to the control.

Pulse 3 is 2 volts. It is accepted because the sum is now 10 volts, and a 1 goes to control. Pulse 4 of 1 volt is rejected, because the sum of 11 would exceed 10.5. So, a 0 is sent to the control. Pulse 5 would bring the sum to 10.8, which is too high, so it, too, is rejected.

Pulse 6 brings the sum to 10.4 volts, which is accepted, sending a 1 to the control. Pulse 7 produces a 0 because the sum 10.6 is too high. Pulse 8 produces a 1 because it brings the sum up to 10.5 volts, the value of the input signal. The 10.5-volt analog signal now is described in digital form as 1010.0101. It is sent from the control to the display, and another sample begins.

Where both the ramp ADC and the dual-slope ADC required up to 1000 pulses for a measurement, the successive-approximation ADC requires only 8. Other versions of this system may require 12 pulses, but even so, this system is much faster than the other two. In addition, rejection of the first pulse (8 volts) could be used to generate an overload command. Likewise, this signal or the acceptance of all pulses could be used to command the FETs in the autoranging input circuitry.

○ Example of a Successive-Approximation ADC

Consider a voltage of 9.3 volts, which is to be measured by a successive-approximation ADC. The pulse number, its voltage value, and the resulting digital output are given in the following table.

Pulse Number	Value Represented	Output
1	8	1
2	4	0
3	2	0
4	1	1
5	0.8	0
6	0.4	0
7	0.2	1
8	0.1	1

The digital representation of the analog voltage of 9.3 volts is 1001.0011.

Displays

Once a voltage has been expressed in digital form, it is ready to be displayed by the digital meter. So, the next step in the measurement process is to send the digital form to a display driver circuit, often part of the LSI chip of the ADC, although we discuss it separately here. The driver first converts the BCD data describing the unknown voltage into decimal digits. The most significant bit (MSB) is on the left end, the least significant bit (LSB) is on the right, and all others are in between.

Next, the driver divides each decimal digit into seven similar, but separate, low-voltage signals. Each group of seven is then sent on to its respective LED or LCD, as indicated in Figure 6–10. Then, each LED or LCD is activated according to which of its seven segments receives a signal and which does not.

Combinations of the seven major display segments are used to produce the decimal numbers from 0 to 9, a minus sign (−), and, occasionally, some letters. The eighth segment is used for a decimal point. Typically, the left (MSB) digit is only used as a 0 or a 1, leading to its being called a 1/2 digit. For this reason, a 4-digit LED meter is often described as having 3 1/2 digits.

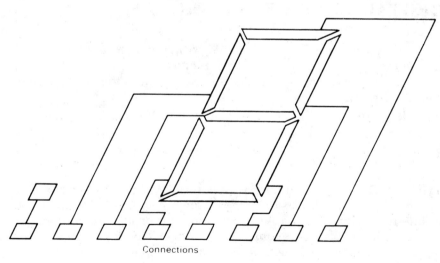

Connections

──o **FIGURE 6–10**
LED/LCD Digit

Whether a meter uses LEDs or LCDs depends on design and application criteria. An LED glows when activated. Red is the most common color used, although green, yellow, white, and other colors are available. A voltage between 1.7 and 3.3 volts and a power between 10 and 100 milliwatts are required for energization. Response time is in nanoseconds, and life expectancy is above 100,000 hours.

An LCD does not glow upon activation. It is normally translucent (clear), but it becomes opaque when it is energized. The liquid crystal container has a transparent front and a reflective back. In daylight, incident light penetrates and reflects out through the crystal, except at the energized segments, where it is absorbed. Thus, a number appears as black on a silver background. When there is an absence of incident light, such as at night, an incandescent lamp is used to back-light the crystal. This light also shows the transparent/opaque contrast.

An LCD operates on microwatts, compared with the milliwatts for LEDs. This advantage is offset by a shorter life expectancy (10,000 hours) and slower speed (a tenth of the speed of an LED). The low power consumption of LCDs makes them popular for hand-held meters.

DIGITAL MULTIMETER CIRCUITS

Most digital voltmeters are included within a multifunction digital multimeter (DMM), such as the one pictured in Figure 6–11. While they are similar to the analog VOM in application, they differ somewhat in circuitry, capability, and operation. Our discussion begins with the multirange AC–DC voltmeter section.

Voltmeter Circuits

Typical DC voltmeter circuits were described earlier in this chapter. In these circuits, voltages to be measured are connected to the input of an attenuator-FET amplifier combination or an operational amplifier with range switch–controlled feedback (see Figure 6–3). One of these arrangements is generally used to bring the unknown voltage to the 0.1- or 0.2-volt maximum level required for the input of the ADC. If the unknown voltage is AC, it must be changed to DC before the measurement can proceed.

As explained in Chapter 5, the AC can be changed to DC with a half-wave or full-wave rectifier. However, as was also explained,

─────○ **FIGURE 6–11**
Digital Multimeter (Photo courtesy of Hickok Electrical Instruments Co., Inc., Cleveland, Ohio)

waveform error occurs whenever the AC signal is not a true sine wave, which quite often is the case. While rectifier circuits are still used, they are becoming less common. True rms circuits, like the circuit in Figure 6–12, are taking their place. A true rms circuit

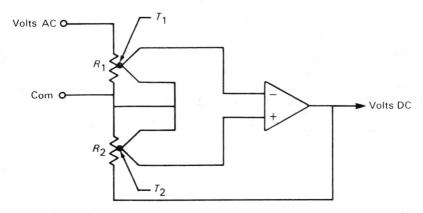

─────○ **FIGURE 6–12**
Thermocouple True rms Circuit

indicates the actual rms, or effective, value of a signal, regardless of its waveshape. Two methods for obtaining true rms in voltmeter circuits will be described here.

With the first method, an AC voltage to be measured is scaled to a workable size and applied across resistance element R_1 (see Figure 6–12). The element warms in response to the effective value of the applied voltage times the current it causes. That is, the temperature rises according to the square of the applied voltage. Thermocouple T_1 detects this change and produces a proportional voltage, which is then amplified. Note that the thermocouple is thermally connected to but electrically insulated from the resistor.

Because the ADC has linear calibration, the AC-to-DC conversion must be changed from nonlinear to linear. This conversion is done by the feedback of a signal from the amplifier output and its application across resistance element R_2. Since the T_1 voltage entered the inverting ($-$) input, the output voltage that is fed back is equal to but the opposite of the input. Thermocouple T_2 responds to the heat developed by this feedback, sending a voltage to the noninverting ($+$) input of the amplifier. Because this amplifier is a differential amplifier, its output is proportional to the difference of the two input signals. It eventually balances at the point where the output voltage produces the same heating effect across R_2 that the input AC voltage produces across R_1.

A second method for determining true rms involves continuous, high-rate calculation. The LSI circuitry squares, calculates the mean of, and then takes the square root of the input signal at a rate of up to 40,000 calculations per second. As with other devices, technical innovation will continue the trend away from mechanical systems and toward microelectronic systems in digital multimeters.

Ammeter Circuits

A common approach to current measurement with digital instruments is to change the current to a voltage and proceed as described for the voltmeter. The circuit in Figure 6–13 is an example of a typical approach. Here, resistors R_1 through R_4 are connected in series with the current to be measured. In low-current situations, all four resistors are included in the shunt. The small voltage produced goes on to the operational amplifier, which, once again, uses the ratio of R_5 and R_6 to control gain.

As the current being measured becomes larger, less resistance is needed in order to have sufficient voltage to measure. Thus, the

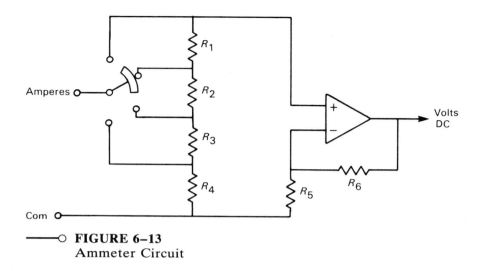

─────○ FIGURE 6–13
Ammeter Circuit

switch is moved down. Moving the switch also reduces I^2R losses and produces the desired lower input impedance.

Ohmmeter Circuits

Like most other ohmmeter instruments, a digital ohmmeter measures resistance by applying voltage across or current through the unknown resistance and processing the response. In the circuit in Figure 6–14A, a voltage is placed across the series combination of an internal standard resistance R_s and the unknown resistance R_x. The ratio of the two voltages is then sent on to the ADC.

Changing ranges also changes the applied voltage, the standard resistor, and the calibration circuit. A typical test voltage depends on the range and is usually less than 3 or 4 volts for the highest and lowest ranges. Middle-range resistances generally use about 1 volt or less.

The system in Figure 6–14B uses a constant-current source. An appropriate and known current level is selected with the resistance range switch and passed through the unknown resistance. The resulting voltage drop is then sent on to the ADC.

This process is designed around Ohm's law; that is, $R = V/I$. Since I is controlled and V is measured, the system is calibrated to display R. Typical test currents are generally around 1 milliampere for the highest and lowest ranges. Middle ranges typically use 100 microamperes or less.

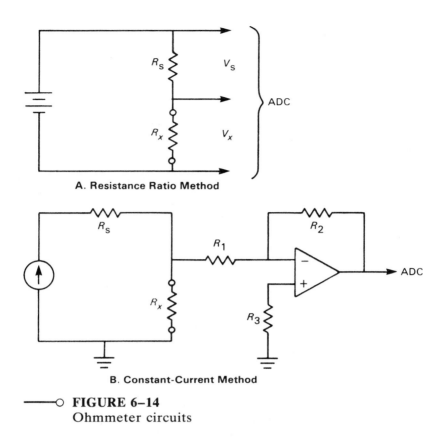

A. Resistance Ratio Method

B. Constant-Current Method

──○ **FIGURE 6–14**
Ohmmeter circuits

DIGITAL MULTIMETER SPECIFICATIONS

Digital VOM manufacturers offer varying descriptions of the capabilities of their instruments. Some will describe five or six capabilities, and others will list dozens. The specifications described here are the ones users are most often interested in and manufacturers are most likely to describe. Remember that capabilities vary greatly among different models. So, always check the manual rather than make incorrect assumptions.

Parameters and Ranges

All digital multimeters measure DC and AC voltage. Most have from four to six ranges, from a low of 0.1 millivolt to a high of 2000 volts.

Most older models averaged AC but indicated rms. Today, about 50% now measure true rms, and the percentage is increasing.

All DMMs also measure DC and AC current. Again, four to six ranges are typical, with a minimum of 0.1 milliampere and a maximum of 2 amperes. All DMMs measure resistance, too, and they generally have four to six ranges. Most common is a low range of 200 ohms and a high of 20 megohms.

Various other features can be found on digital VOMs. Diode check is very common, as is decibel measurement. Power measurement is available on some, and others can be used with a temperature probe.

Accuracy

While analog meter accuracy is almost always a percentage of full scale, digital meters very seldom use full-scale percentages. Most of the time, digital meter accuracy is specified as a percentage of the *indicated value.* For DC voltage, accuracy is usually between 0.05% and 0.1%, with 0.001% available on more expensive models. Most DMM accuracy specifications include the additional comment " ±1 digit" because of the rounding-off process.

Accuracy for DC current is typically from 0.5% to 1%, and resistance accuracy is from 0.2% to 1%. Accuracy specifications for AC voltage and current include frequency ranges, which will be discussed next. The ability of a meter to accurately measure pulses can be determined by the *crest factor* specification. Crest factor is equal to the peak voltage divided by the rms voltage of a given signal. It describes, mathematically, the variability in one cycle of a signal. Specifications of 5 : 1 are typical.

Frequency Response

The frequency range over which AC voltage and current accuracy is defined varies greatly from one meter to the next. If there is a weak link in the attributes of digital meters, this specification is it.

As an example, a specification may state 0.5% accuracy between 50 and 500 hertz but 1% between 30 and 50 hertz and between 500 and 2000 hertz. Quite often, accuracy deteriorates above 1000 to 5000 hertz. Sometimes, deterioration begins at 400 or 500 hertz. Always know the frequency response of a digital multimeter before you measure a signal above or below line frequency.

Display

The most common displays used with portable instruments are LCDs. Some bench-top instruments use LEDs. Most portable instruments have 3 1/2 or 4 1/2 digits. The more expensive, higher-accuracy, bench-top models have more digits, often up to 7 1/2. As explained earlier, the 1/2 digit is the left digit (MSB), and it can represent only 0 or 1.

Sensitivity and Protection

An input impedance of 10 megohms on all voltage ranges is a typical sensitivity specification for most digital multimeters. Most DMMs have protection from input overload. Circuit breakers, fuses, auto-ranging, and diode clipper circuits are some of the methods used. In addition, solid-state circuitry and displays make the digital multimeter much more rugged than its analog counterpart. In fact, drop test tolerance is often a specification. The ability to survive a 10-foot drop is quite common.

Accessories

Manufacturers offer a variety of accessories for DMMs, such as cases, thermocouples, and assorted probes. For example, a current transformer can be used to measure current of 600 amperes or more by clamping the transformer around the conductor without a circuit disconnection. This modification produces a voltage that can be measured with a DMM and translated into current magnitude with calibration data.

High-voltage probes allow the user to measure voltages normally beyond the range of the DMM. Values of 5000 and 6000 volts are typical, although 30,000 and 40,000 are available. Radio frequency probes permit the user to measure small AC voltages to 500 megahertz and beyond.

DIGITAL MULTIMETER APPLICATIONS

Digital multimeters are typically used to measure voltage, current, and resistance and to check diodes and circuit continuity. Analog VOM operation was explained in detail in Chapter 5, and digital

VOM operation is very similar. So, DMM operation will merely be reviewed here, with differences noted. However, you should always read the manual, especially about how to check whether the meter is working properly (for instance, battery life).

Voltage Measurement

The following sequence is suggested for voltage measurement with a DMM:

1. Connect the leads to the instrument.
2. Select volts and DC or AC.
3. Select the highest range.
4. Zero the meter if it does not have automatic zeroing.
5. Connect the ground or common lead to the circuit being measured.
6. Connect the positive lead to the circuit.
7. Energize the circuit.
8. Switch down the range to the lowest range that does not exceed the display limits. Most meters have an overload indicator.
9. Recheck zero.
10. Read the indicated value.
11. Disconnect the leads from the circuit, and switch the meter off.

Current Measurement

The following steps can be used for current measurement with a DMM:

1. Connect the leads to the instrument.
2. Select amperes or milliamperes and the highest range.
3. Zero the meter, if necessary.
4. De-energize the circuit to be measured.
5. Disconnect the circuit at the point where current is to be measured.
6. Connect the instrument probes to the separated wires, with minus toward the negative part of the circuit and plus toward the positive part.
7. Energize the circuit.
8. Switch down the range to the lowest range that does not exceed the display limits.

9. Recheck zero.
10. Read the indicated value, and de-energize the circuit.
11. Disconnect the probes, and reconnect the circuit.
12. Switch the meter off.

Resistance Measurement

The resistance of a component should not be measured if the component is in a circuit or if the circuit is alive. De-energize the circuit, and disconnect at least one lead of the component from that circuit. Then, perform the following steps for resistance measurement:

1. Select the ohms function and the highest range.
2. Zero the meter, if necessary.
3. Connect the probes across the unknown resistance, being sure that your fingers do not touch either probe tip.
4. Switch ranges until you find the range giving the highest reading without display overload.
5. Recheck zero.
6. Read the value from the display.
7. Disconnect the probes, and reconnect the circuit.
8. Switch the meter off.

Diode Check

Many digital VOMs have a resistance range for use in checking diodes. With it, you can check the front-to-back ratio of a diode or a junction transistor. Usually, this range uses a voltage slightly over 1 volt, enough to exceed a junction's forward-conduction threshold.

To check a diode, measure the resistance first in one direction and then in the other. Overrange is indicated in the reverse-biased direction, and a low resistance (see the specific manual) is indicated in the forward direction. Since there is such a variety of ohmmeter circuits, it is best to read the manual first so that you will know the semiconductor-checking capabilities and procedures for your DMM.

SUMMARY

A DC voltage coming into a digital voltmeter for measurement is first scaled up or down to a workable magnitude for the stage that follows. This signal preparation is usually done by an attenuator followed by a FET amplifier or by an operational amplifier with range switch–controlled feedback. This stage also establishes the high input impedance seen by the source, typically 10 megohms. In some circuits, an autoranging system controls the operational amplifier gain with feedback from later stages.

The properly scaled DC voltage then proceeds to an analog-to-digital converter (ADC), where it is converted to a binary-coded decimal (BCD) expression. The ADC circuit is usually contained on a large-scale integration (LSI) chip. Three common ADC systems are in use.

One converter system is the ramp ADC. It produces a linear DC voltage, which is compared with 0 volts and the unknown voltage. Comparators control a counter, which counts the pulses from an oscillator. The ramp ADC is quite accurate because it is designed around a crystal-controlled oscillator, a DC voltage standard, and a counter, all of which can be made with accuracy and high quality.

The second system is called the dual-slope ADC. In it, the unknown voltage charges a capacitor for a controlled period of time. Then, discharge time through a constant-current source is measured. This system is less sensitive to transients than the ramp ADC.

The third system is the successive-approximation ADC. It takes a brief sample of the unknown voltage and checks it against the sum of a series of decreasing standard signals. For example, a 10.40-volt incoming signal will be checked against 8-, 4-, 2-, 1-, 0.8-, 0.4-, 0.2-, and 0.1-volt signals. This checking system would produce a digital value of 1010.0100.

Both LEDs and LCDs are used to display measured values. The LEDs are used on many bench-top instruments, and the LCDs are used for most portable models because of their lower power consumption. Displays receive their power from a display driver circuit between them and the ADC.

In digital multimeters, a DC voltage is measured as just described. An AC voltage must be changed to DC before it can be measured, as in analog meters. Some meters use rectifiers to produce an average value, which is then measured. True rms meters measure the actual value by using a thermocouple system or by using a high-speed system that squares, averages, and finds the square root of the input signal.

Current is measured in a DMM by passing the current through an internal shunt of known resistance and producing a voltage. This voltage is then measured and described in terms of current. Resistance is measured by passing a known current through the unknown resistance and measuring voltage. Resistance is then internally calculated with Ohm's law.

The typical DMM measures DC voltages from 0.1 to 2000 volts at an accuracy of 0.05% to 0.1%. The AC voltage is measured over the same range, but the accuracy depends on frequency response. The frequency range between 50 and 500 hertz is often the best, with an accuracy of 0.5%. Below 50 and above 500 hertz, the accuracy may diminish to 1% or even more. Frequency response is the digital voltmeter's weakest feature.

In a DMM, typical current ranges are from 0.1 milliampere to 2 amperes, and accuracy is between 0.5% and 1%. Resistance ranges extend from 200 ohms to 20 megohms, and accuracy is from 0.2% to 1%. In most cases, accuracy is stated in terms of ±1 digit and a percentage of indicated value, not full-scale value.

With a DMM, voltage is measured by selecting the highest range, connecting the meter across the circuit, switching down to the lowest usable range, reading the indicated value, and disconnecting the meter. Current is measured by selecting the highest range, breaking the circuit, connecting the meter in series with the break, switching down to the lowest usable range, reading the indicated value, and reconnecting the circuit.

Resistance is measured by removing the component from the circuit, selecting the highest resistance range, connecting the meter across the component, switching ranges to find the best range, reading the indicated value, and reconnecting the circuit. Diodes can be checked by measuring their front-to-back ratio on the resistance scale designated for such tests.

BIBLIOGRAPHY

Focus on benchtop DMM's: performance up, cost down, B. Milne, Electronic Design, April 28, 1983.

Hand-held DMM using multiple slope techniques, M. Hickman, Electronic Engineering, November, 1983.

Hand-held 3 1/2 digit DMM's feature bar-graph displays, EDN, December 22, 1983.

Digital Electronics, Logic and Systems, 2nd ed., John D. Kershaw, Breton Publishers, 1983.

QUESTIONS

1. Sketch a block diagram of a digital voltmeter, and describe the function of each section.

2. Describe two concerns in signal preparation.

3. Explain the function of autoranging, and describe one method by which it can be accomplished.

4. What is the purpose of an ADC?

5. Describe three ways analog-to-digital conversion can be accomplished.

6. Which ADC method is the simplest, and which method is the fastest?

7. Describe how a digital multimeter measures current.

8. Describe how a digital multimeter measures resistance.

9. Explain the difference between *average* and *effective* values of AC voltage.

10. Describe two methods by which effective values can be obtained.

11. Compare LEDs and LCDs.

12. Which display is most commonly used? Why?

13. Which VOM is most rugged, analog or digital? Why?

14. Name and describe five common specifications of digital multimeters.

15. Explain what is meant by a 3 1/2 digit display.

16. Discuss the advantages of digital multimeters over analog multimeters. Discuss the disadvantages.

17. Describe the differences between accuracy as a percentage of full-scale value and as a percentage of indicated value.

18. Explain what is meant by ± 1 digit.

PROBLEMS

1. Redraw the circuit in Figure 6–3A and perform the necessary calculations to add a 1000-volt range.

2. Calculate the attenuator values required for the circuit in Figure 6–3A if the FET input signal is to remain at 0.1 volt for full-scale inputs of 0.1, 0.5, 1, 5, and 10 volts.

3. Calculate the attenuator values required for the circuit in Figure 6–3A if the FET input signal is to remain at 0.5 volt for full-scale inputs of 0.5, 1, 5, 10, 50, and 100 volts.

4. Calculate the values needed for R_{2a}, R_{2b}, R_{2c}, and R_{2d} of the circuit in Figure 6–3B in order to produce ranges of 0.1, 1, 5, and 10 volts, full-scale. Assume that R_1 is 100 kilohms and that a full-scale output voltage from the amplifier is 1 volt.

5. Repeat Problem 4 for producing ranges of 1, 5, 10, and 50 volts.

6. Repeat Problem 5 when the full-scale output of the amplifier is 0.1 volt.

7. For the system in Figure 6–9, what analog value would be represented by the digital value 1101.0010?

8. For the system in Figure 6–9, what analog value would be represented by the digital value 0101.0110?

9. Using the method illustrated in Figure 6–9, determine the digital expression for 11.3 volts.

10. How would 3.2 volts be expressed by the system in Figure 6–9? How would it be expressed if automatic shift occurred to make the first digit 1?

11. Suppose a digital meter has an accuracy of 1% ±1 digit. Between what limits is the true value of a voltage indicated as 1.352 volts?

12. Repeat Problem 11 for a meter with an accuracy of 0.1% ±1 digit.

13. Suppose a digital meter has an accuracy of 2% ±1 digit. Between what limits is the true value of a voltage indicated as 73.5 volts?

14. Repeat Problem 13 for a meter with an accuracy of 0.5% ±1 digit.

7

Bridges

OBJECTIVES

This chapter describes some common bridge circuits, their applications, and their specifications. Traditional manual measurement systems and more recent automatic systems are also described.

Discuss the differences between a meter and a bridge type of measurement instrument.
Identify parameters that can be measured with bridges.
Name and describe the operation of a basic DC bridge circuit.
Name and describe the operation of three basic AC bridge circuits.
Describe how to measure resistance with a DC bridge.
Describe how to measure inductance, quality factor, capacitance, and dissipation factor with an AC bridge.

DC BRIDGE CIRCUITS

Bridges have long been used for the measurement of resistance R, inductance L, and capacitance C. While various AC bridges are used for the L and C measurements, one DC bridge is traditionally used for the measurement of R. The discussion of the DC circuit in this section begins with an introduction to balanced and unbalanced bridges.

Bridge versus Meter

A VOM measures resistance by applying a DC voltage across that resistance and responding to the amount of current that flows. The current deflects a pointer upscale, which, in turn, indicates a resistance value. This system is open-ended and has an accuracy that depends on the quality of the internal resistors and the meter movement. In addition, the accuracy of the assumed value depends on operator interpretation, a significant concern with analog VOMs.

Two characteristics of bridge circuits enhance their accuracy. First, bridges are closed systems that operate on the principle of measurement by comparison. That is, *a known (standard) value is adjusted until it is equal to an unknown value.* This process of adjusting for equality is called *null balance.* The word *null* used here means "no difference." With two values equal and one of them known to a high degree of accuracy, the other value is then also known.

A second advantage of bridge measurement systems results from the null-balance approach. While a meter is part of this system, it is not used to measure a specific magnitude. Instead, it is used to indicate the null condition. Therefore, its accuracy and its ease of interpretation are no longer significant factors. All the meter—and the operator—must do is indicate and recognize the null condition.

Wheatstone Bridge

Our introduction to the *Wheatstone bridge,* the DC bridge for measuring resistance, begins with a review of DC series-parallel circuits. Consider the circuit in Figure 7–1A. Recall from your study of series-parallel circuits that the voltage at point *A* from common (or negative) is 6 volts. It is 6 volts because a total of 12 volts is applied and the series resistors are equal, producing equal voltage drops. The voltage at point *B* is also 6 volts, for the same reason.

In Figure 7–1B, the voltage at point *A* is 8 volts because the 2-ohm resistor drops twice the amount of voltage as the 1-ohm resistor drops. Again, points *A* and *B* have the same voltage. As you will see shortly, our concern is the difference between these two voltages. Since these voltages are equal, this bridge is said to be *balanced.*

The bridge in Figure 7–1C is also balanced, because the voltages at *A* and *B* are equal. The fact that one combination is 1 and 2 ohms while the other is 10 and 20 ohms should also be noted. Of most importance is the difference between voltages *A* and *B,* not their specific values.

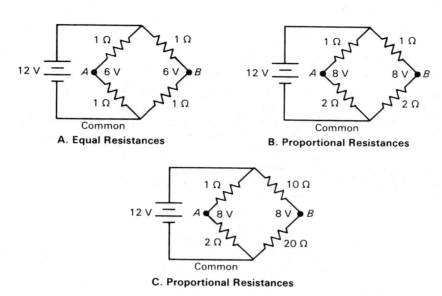

FIGURE 7–1
Balanced Bridges

The bridge in Figure 7–2A has 8 volts at point A and 6 volts at point B. In Figure 7–2B, the right-branch ratio produces 10 volts at point B. In Figure 7–2C, again, notice that there is a 2-volt difference between points A and B, with B more positive than A. These bridges are said to be *unbalanced*.

In summary, a bridge is balanced when the voltages at points A and B are equal. It is unbalanced when the voltages are unequal. The specific magnitudes of the voltages are not as significant as their ratios, as we will see shortly.

Figure 7–3 shows the basic Wheatstone bridge circuit. Balance is achieved when the ratio of R_2 to R_1 is the same as the ratio of R_x to R_3. Stated mathematically, the condition for balance is as follows:

$$\frac{R_2}{R_1} = \frac{R_x}{R_3} \quad \text{or} \quad R_x = \frac{R_2 \times R_3}{R_1}$$

Resistors R_1 and R_2 are called *ratio resistors*. They are precision fixed resistors and can be connected or changed with a range switch. Resistor R_3 is a precision rheostat with a circumference between 15 and 18 inches. This length provides far greater resolution or precision than the more common 1-inch-diameter potentiometer. It is connected to a calibrated dial on the front of the instrument, which

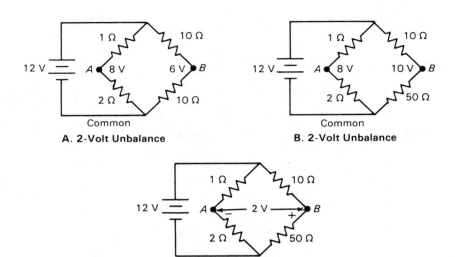

A. 2-Volt Unbalance

B. 2-Volt Unbalance

C. Polarity of 2-Volt Unbalance

——○ **FIGURE 7–2**
Unbalanced Bridges

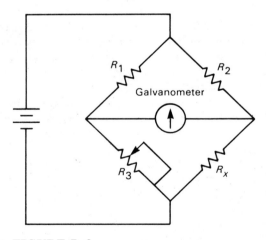

——○ **FIGURE 7–3**
Wheatstone Bridge

is used for the indication of component values. Generally, internal batteries power the bridge. A panel-mounted, zero-centered galvanometer is used as the *null detector*. While component values may vary from one bridge to another, all Wheatstone bridges operate as described in the next subsection.

Measurement of Resistance

Imagine that the ratio resistors and the rheostat R_3 of a Wheatstone bridge are 1000 ohms each. In addition, a calibrated dial is attached to the rheostat to indicate the value to which the rheostat is adjusted. The unknown resistance R_x is connected as shown in Figure 7–4A. Power is applied, and R_3 is adjusted until the meter indicates null. Then, the dial indication is read. The example that follows illustrates the calculations for resistance measurements.

○ Example of a Wheatstone Bridge

Assume, as shown in Figure 7–4A, that balance occurs when the dial indicates 500 ohms. The unknown must also be 500 ohms, for that is the only value that would produce balance. Thus, we have the following calculation:

$$R_x = \frac{R_2 \times R_3}{R_1} = \frac{1000 \, \Omega \times 500 \, \Omega}{1000 \, \Omega} = 500 \, \Omega$$

Unfortunately, this arrangement would only work for resistances up to 1000 ohms, because that is the maximum limit of the rheostat. For measurement of resistances above 1000 ohms, other resistors are switched in. For example, changing R_2 to 10,000 ohms extends the range upward to 10,000 ohms. As indicated in Figure 7–4B, an R_x of 5000 ohms can be balanced with R_3 on midrange if R_2 is 10,000 ohms. The calculation is as follows:

$$R_x = \frac{R_2 \times R_3}{R_1} = \frac{10,000 \, \Omega \times 500 \, \Omega}{1000 \, \Omega} = 5000 \, \Omega$$

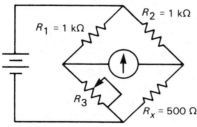

A. Balanced When
R_3 at 500 Ω for R_2 = 1 kΩ

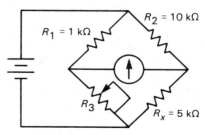

B. Balanced When
R_3 at 500 Ω for R_2 = 10 kΩ

——○ **FIGURE 7–4**
Measuring Resistance with a Wheatstone Bridge

The range switch of a Wheatstone bridge can generally connect one of six or eight different precision values. The operator needs only to follow these steps:

1. Connect the unknown resistance.
2. Select the best range.
3. Adjust the rheostat for null.
4. Read the dial indication.
5. Multiply the reading by the range multiplier.

○ Example of Calculation for a Wheatstone Bridge Measurement

Consider a Wheatstone bridge with the following values: R_1 = 500 ohms, R_2 = 600 ohms, and R_3 = 100 ohms. Then, the unknown resistance R_x is calculated as follows:

$$R_x = \frac{R_2 \times R_3}{R_1} = \frac{600\ \Omega \times 100\ \Omega}{500\ \Omega} = 120\ \Omega$$

AC BRIDGE CIRCUITS

Inductance and capacitance of AC circuits can also be quite accurately measured with bridge instruments. As with resistance, the null balance and a comparison of unknowns with knowns are used. The difference here is that reactance rather than resistance is the primary parameter used for comparison.

Maxwell Bridge

Two significant differences exist between inductance or capacitance bridges and resistance bridges. First, the opposition from a resistor to AC is essentially the same as it is to DC, although its reactive components can become significant at RF. Inductive and capacitance reactances, however, are parameters that are highly dependent on frequency. Second, inductors and capacitors have resistance components whose values are *not* negligible and, hence, cannot be ignored. For these reasons, inductance and capacitance bridges are AC-operated and can measure the effects of resistance in inductors and capacitors.

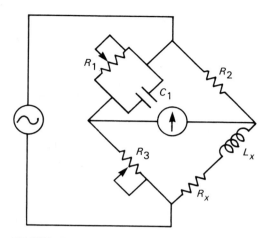

—○ **FIGURE 7–5**
Maxwell Bridge

While there are many different inductance (L) and capacitance (C) bridge circuits, the *Maxwell bridge circuit* has long been popular. It is used for the measurement of the inductance of components with a quality factor Q of less than 10. As with the Wheatstone bridge, a voltage is placed across the bridge, and an adjustable rheostat is varied until the null condition is achieved.

Since the reactance of the unknown inductance is being balanced with the standard resistances, AC is placed across the bridge. Furthermore, since reactance is frequency-dependent, the applied frequency must be accurately known and controlled. Most AC bridges are calibrated for use with an internal, 1-kilohertz voltage source, with the option of using an external generator. Headphones or an oscilloscope can be used as an externally connected detector instead of the internally connected meter.

Figure 7–5 shows the circuit for a Maxwell bridge. As indicated in the figure, a standard rheostat R_3 is used to match the reactance X_L of the unknown inductance L_x. Rheostat R_1 balances the winding resistance of the typical low-Q inductor.

In a multibridge instrument (or impedance bridge, to be described shortly), the Maxwell bridge is selected with the L_s mode. Here, the inductor has a low Q influenced by series resistance. To balance this bridge, the operator alternately adjusts R_1 and R_3 until null is achieved. Then, two values are read from the dials and the range switches. The main rheostat R_3 indicates the units of inductance. Rheostat R_1 indicates the quality (or storage) factor Q, where Q is defined, mathematically, as follows:

$$Q = \frac{X_L}{R_x}$$

where R_x in the Maxwell bridge equals the series resistance.

Another way to consider this circuit is to assume that R_3 balances the imaginary component X_L of the opposition, while R_1 balances the series component R_x of the opposition. As with other bridges, R_2 and C_1 can be changed with range switches in order to produce ratios that will balance a wide range of inductances.

The equations for a balanced Maxwell bridge are as follows:

$$L_x = R_2 \times R_3 \times C_1$$

$$R_x = \frac{R_2 \times R_3}{R_1}$$

○ Example of Calculations for a Maxwell Bridge

Consider a Maxwell bridge with the following component values: $R_1 = 470$ kilohms, $R_2 = 47$ kilohms, $R_3 = 4.7$ kilohms, and $C_1 = 0.1$ microfarad. The calculations for this bridge are as follows:

$$L_x = R_2 R_3 C_1 = (47 \times 10^3 \, \Omega) \times (4.7 \times 10^3 \, \Omega) \times (0.1 \times 10^{-6} \, F)$$
$$= 22.09 \, H$$

$$R_x = \frac{R_2 R_3}{R_1} = \frac{(47 \times 10^3 \, \Omega) \times (4.7 \times 10^3 \, \Omega)}{470 \times 10^3 \, \Omega} = 470 \, \Omega$$

○ Example of Calculations for Another Maxwell Bridge

Consider a Maxwell bridge with the following values: $R_1 = 1$ megohm, $R_2 = 50$ kilohms, $R_3 = 10$ kilohms, and $C_1 = 0.01$ microfarad. Then, the unknown inductance L_x and resistance R_x are calculated as follows:

$$L_x = R_2 R_3 C_1 = (50 \times 10^3 \, \Omega) \times (10 \times 10^3 \, \Omega) \times (0.01 \times 10^{-6} \, F)$$
$$= 5 \, H$$

$$R_x = \frac{R_2 R_3}{R_1} = \frac{(50 \times 10^3 \, \Omega) \times (10 \times 10^3 \, \Omega)}{1 \times 10^6 \, \Omega} = 500 \, \Omega$$

Hay Bridge

The *Hay bridge,* whose circuit is shown in Figure 7–6, is used for the measurement of inductances having a Q between 10 and 100.

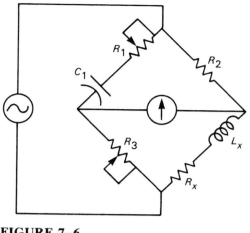

────○ **FIGURE 7–6**
Hay Bridge

Here, the inductance measured is assumed to have a high parallel resistive component rather than the low series component of the Maxwell bridge. Where the mode was L_s for the Maxwell bridge, it is L_p for the Hay bridge. The value of Q is defined differently as well:

$$Q = \frac{R_x}{X_L}$$

where R_x in the Hay bridge equals the parallel resistance.

One difference between the Maxwell and the Hay circuits involves the relationship of R_1 and C_1. They are connected in parallel in the Maxwell bridge, and they are connected in series in the Hay bridge. Low-Q circuits measured by the Maxwell bridge can be easily balanced by moderate-sized R_1 values. However, the large parallel R_1 values needed to balance a high Q could cause leakage problems. Therefore, the series R_1 and C_1 circuit of the Hay bridge is more appropriate for high-Q inductance measurements.

Two additional points about these circuits should be noted. First, an adjacent resistor R_3 is used to measure inductance, rather than an adjustable inductor, because high-quality, linear, adjustable resistors are far easier to produce than inductors with the same features. Second, a capacitor C_1 positioned diagonally opposite the unknown inductance is used rather than an adjacent inductor because the accuracy and stability required in a reactive component are easier to achieve with capacitors than with inductors.

Operator procedure is the same with both the Maxwell and the Hay bridge. The main rheostat R_3 is used to balance the reactive component of the inductor. Rheostat R_1 balances the parallel resistance. Rheostat R_3 indicates inductance, while R_1 indicates quality factor.

The equations for a balanced Hay bridge are as follows:

$$L_x = \frac{R_2 R_3 C_1}{1 + (2\pi f)^2 C_1^2 R_1^2}$$

$$R_x = \frac{(2\pi f)^2 C_1^2 R_1 R_2 R_3}{1 + (2\pi f)^2 C_1^2 R_1^2}$$

In these equations, f is the symbol for frequency, and π (Greek letter pi) represents a constant equal to 3.14 (rounded off).

○ Example of Calculations for a Hay Bridge

Consider a Hay bridge with the following components: $R_1 = 200$ ohms, $R_2 = 1$ kilohm, $R_3 = 10$ kilohms, $C_1 = 0.1$ microfarad, and $f = 1$ kilohertz. The calculations for this bridge are as follows:

$$L_x = \frac{R_2 R_3 C_1}{1 + (2\pi f)^2 C_1^2 R_1^2}$$

$$= \frac{(1 \times 10^3\,\Omega) \times (10 \times 10^3\,\Omega) \times (0.1 \times 10^{-6}\,\text{F})}{1 + (2\pi \times 1 \times 10^3\,\text{Hz})^2 \times (0.1 \times 10^{-6}\,\text{F})^2 \times (2 \times 10^2\,\Omega)^2}$$

$$= \frac{1}{1 + (39.48 \times 10^6 \times 0.01 \times 10^{-12} \times 4 \times 10^4)}\,\text{H}$$

$$= 0.984\,\text{H}$$

$$R_x = \frac{(2\pi f)^2 C_1^2 R_1 R_2 R_3}{1 + (2\pi f)^2 C_1^2 R_1^2}$$

$$= \frac{(2\pi \times 1 \times 10^3\,\text{Hz})^2 \times (0.1 \times 10^{-6}\,\text{F})^2 \times (2 \times 10^2\,\Omega) \times (1 \times 10^3\,\Omega) \times (10 \times 10^3\,\Omega)}{1 + (2\pi \times 1 \times 10^3\,\text{Hz})^2 \times (0.1 \times 10^{-6}\,\text{F})^2 \times (2 \times 10^2\,\Omega)^2}$$

$$= \frac{39.48 \times 10^6 \times 1 \times 10^{-14} \times 2 \times 10^2 \times 1 \times 10^3 \times 1 \times 10^4}{1 + (39.48 \times 10^6 \times 0.01 \times 10^{-12} \times 4 \times 10^4)}\,\Omega$$

$$= 802\,\Omega$$

○ Example of Calculations for Another Hay Bridge

Consider the Hay bridge with the following values: $R_1 = 100$ ohms, $R_2 = 1$ kilohm, $R_3 = 10$ kilohms, $C_1 = 0.2$ microfarad, and $f = 1$ kilohertz. The calculations for the unknowns in this bridge are as follows:

$$L_x = \frac{R_2 R_3 C_1}{1 + (2\pi f)^2 C_1^2 R_1^2}$$

$$= \frac{(1 \times 10^3 \,\Omega) \times (10 \times 10^3 \,\Omega) \times (0.2 \times 10^{-6} \,\text{F})}{1 + (2\pi \times 1 \times 10^3 \,\text{Hz})^2 \times (0.2 \times 10^{-6} \,\text{F})^2 \times (1 \times 10^2 \,\Omega)^2}$$

$$= \frac{2}{1 + (39.48 \times 10^6 \times 0.04 \times 10^{-12} \times 1 \times 10^4)} \,\text{H}$$

$$= 1.97 \,\text{H}$$

$$R_x = \frac{(2\pi f)^2 C_1^2 R_1 R_2 R_3}{1 + (2\pi f)^2 C_1^2 R_1^2}$$

$$= \frac{(2\pi \times 1 \times 10^3 \,\text{Hz})^2 \times (0.2 \times 10^{-6} \,\text{F})^2 \times (1 \times 10^2 \,\Omega) \times (1 \times 10^3 \,\Omega) \times (10 \times 10^3 \,\Omega)}{1 + (2\pi \times 1 \times 10^3 \,\text{Hz})^2 \times (0.2 \times 10^{-6} \,\text{F})^2 \times (1 \times 10^2 \,\Omega)^2}$$

$$= \frac{39.48 \times 10^6 \times 0.04 \times 10^{-12} \times 1 \times 10^2 \times 1 \times 10^3 \times 10 \times 10^3}{1 + (39.48 \times 10^6 \times 0.04 \times 10^{-12} \times 1 \times 10^4)} \,\Omega$$

$$= 155 \,\Omega$$

Typically, the Maxwell and Hay bridges are part of a multicircuit impedance bridge, which will be described shortly. The appropriate bridge is not selected by name but by the quality factor Q of the inductance. Since that value is probably not known at this point, the bridge choice can only be a guess. However, the bridge will not balance if the wrong bridge is tried.

Schering Bridge

A *Schering bridge* balances with component capacitance and its series resistance. It indicates capacitance C and dissipation factor D. The dissipation factor is defined as follows:

$$D = \frac{R_x}{X_C}$$

In this equation, X_C is the reactance due to the capacitance.

Figure 7–7 shows the basic circuit for the Schering bridge, which is somewhat different from the inductance bridges described earlier. Capacitor C_1 is a standard capacitor, selected with a range switch and producing an opposition ratio with R_2 that balances with the unknown capacitance C_x. Rheostat R_2 is calibrated in units of capacitance and is used in conjunction with C_2 for reactive balance. The combination of C_2 and R_1 balances the resistive component of the capacitor. The adjustable capacitor C_2 is calibrated to indicate the dissipation factor D of the capacitor; D, like Q, has no units.

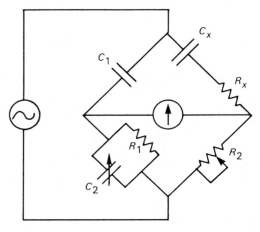

──○ **FIGURE 7–7**
Schering Bridge

The equations for a balanced Schering bridge are as follows:

$$C_x = \frac{R_1 \times C_1}{R_2}$$

$$R_x = \frac{R_2 \times C_2}{C_1}$$

○ Example of Calculations for a Schering Bridge

Consider a balanced Schering bridge with the following components: $R_1 = 100$ kilohms, $R_2 = 10$ kilohms, $C_1 = 0.001$ microfarad, and $C_2 = 15$ picofarads. The calculations for this bridge are as follows:

$$C_x = \frac{R_1 C_1}{R_2} = \frac{(100 \times 10^3 \,\Omega) \times (0.001 \times 10^{-6}\,\text{F})}{10 \times 10^3 \,\Omega}$$

$$= \frac{1 \times 10^5 \times 1 \times 10^{-9}}{1 \times 10^4} \,\mu\text{F} = 1 \times 10^{-8}\,\mu\text{F} = 0.01\,\mu\text{F}$$

$$R_x = \frac{R_2 C_2}{C_1} = \frac{(10 \times 10^3 \,\Omega) \times (15 \times 10^{-12}\,\text{F})}{0.001 \times 10^{-6}\,\text{F}}$$

$$= \frac{1 \times 10^4 \times 1.5 \times 10^{-11}}{1 \times 10^{-9}} \,\Omega = 1.5 \times 10^2 \,\Omega = 150\,\Omega$$

○ **Example of Calculations for Another Schering Bridge**

Consider another Schering bridge with the following values: R_1 = 50 kilohms, R_2 = 10 kilohms, C_1 = 0.05 microfarad, and C_2 = 100 picofarads. The unknowns for this bridge are calculated as follows:

$$C_x = \frac{R_1 C_1}{R_2} = \frac{(50 \times 10^3 \, \Omega) \times (0.05 \times 10^{-6} \, F)}{10 \times 10^3 \, \Omega} = \frac{2.5 \times 10^{-3}}{1 \times 10^4} \, \mu F$$

$$= 0.25 \, \mu F$$

$$R_x = \frac{R_2 C_2}{C_1} = \frac{(10 \times 10^3 \, \Omega) \times (100 \times 10^{-12} \, F)}{0.05 \times 10^{-6} \, F} = \frac{1000 \times 10^{-9}}{5 \times 10^{-8}} \, \Omega$$

$$= 20 \, \Omega$$

The capacitance comparison bridge in Figure 7–8 is often used in commercial impedance bridges. Where the Schering bridge has two capacitors, one adjustable and one fixed, the comparison bridge has one fixed standard capacitor. The balance equations for this bridge are as follows:

$$C_x = \frac{R_1 \times C_1}{R_2}$$

$$R_x = \frac{R_2 \times R_3}{R_1}$$

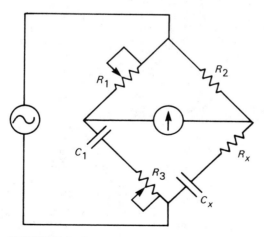

———○ **FIGURE 7–8**
Capacitance Comparison Bridge

○ **Example of Calculations for a Capacitance Comparison Bridge**

Consider a capacitance comparison bridge that balances under these conditions: R_1 = 8 kilohms, R_2 = 10 kilohms, R_3 = 100 ohms, and C_1 = 0.4 microfarad. The calculations for this bridge are as follows:

$$C_x = \frac{R_1 C_1}{R_2} = \frac{(8 \times 10^3 \, \Omega) \times (0.4 \times 10^{-6} \, \text{F})}{10 \times 10^3 \, \Omega} = \frac{3.2 \times 10^{-3}}{10 \times 10^3} \, \mu\text{F}$$

$$= 3.2 \, \mu\text{F}$$

$$R_x = \frac{R_2 R_3}{R_1} = \frac{(10 \times 10^3 \, \Omega) \times (1 \times 10^2 \, \Omega)}{8 \times 10^3 \, \Omega} = \frac{10 \times 10^5}{8 \times 10^3} \, \Omega = 125 \, \Omega$$

○ **Example of Calculations for Another Capacitance Comparison Bridge**

Consider a capacitance comparison bridge with the following components: R_1 = 5 kilohms, R_2 = 10 kilohms, R_3 = 400 ohms, and C_1 = 0.75 microfarad. The calculations for the unknowns in this bridge are as follows:

$$C_x = \frac{R_1 C_1}{R_2} = \frac{(5 \times 10^3 \, \Omega) \times (0.75 \times 10^{-6} \, \text{F})}{10 \times 10^3 \, \Omega} = \frac{3.75 \times 10^{-3}}{10 \times 10^3} \, \mu\text{F}$$

$$= 0.375 \, \mu\text{F}$$

$$R_x = \frac{R_2 R_3}{R_1} = \frac{(10 \times 10^3 \, \Omega) \times (4 \times 10^2 \, \Omega)}{5 \times 10^3 \, \Omega} = \frac{40 \times 10^5}{5 \times 10^3} \, \Omega = 800 \, \Omega$$

IMPEDANCE BRIDGES

Almost any laboratory you will work in will have at least one *impedance bridge* such as the bridge shown in Figure 7–9. With the impedance bridge, you can measure, *R, L, C, Q,* or *D* of discrete, passive components. It utilizes circuits such as those described earlier in the chapter and other circuits, assembled into a high-quality, multifunctional package.

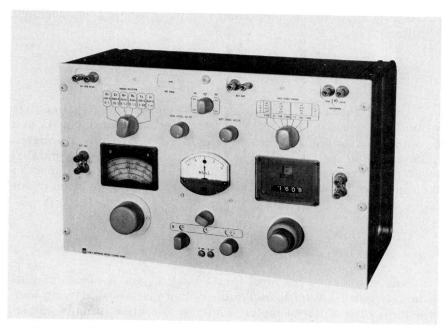

─────○ **FIGURE 7–9**
Impedance Bridge (Photo courtesy of GenRad, Inc.)

Characteristics

The typical impedance bridge performs its measurements accurately and precisely by utilizing the most appropriate of its internal bridge circuits. With it, you can measure parameters such as the DC resistance of a resistor. You can also measure the AC resistance by applying an internal, 1-kilohertz signal or an alternative, externally connected, AF voltage. This measurement indicates the *total* opposition a resistor offers in an AC application.

Inductance L and quality factor Q can also be measured with an impedance bridge. Once again, the most appropriate circuit is automatically selected when you choose the L_s or L_p mode. In the L_s mode, component inductance and a low resistance are perceived as being in series, and a Maxwell bridge is connected. In the L_p mode, inductance and a high resistance are perceived as being in parallel, and a Hay bridge is used. Both configurations produce accurate indications of the component's L and Q, where Q is calculated as follows:

$$Q = \frac{X_L}{R_s} = \frac{R_p}{X_L}$$

Capacitance and dissipation factor D measurements also utilize various bridge arrangements. For capacitance measurement, C_s and C_p modes are available. With the C_s mode, a capacitor is considered to have a small series resistance. The C_p mode is used when parallel resistance (leakage) is to be measured. In either case, the impact of the resistance is measured and defined in terms of dissipation factor D:

$$D = \frac{R_s}{X_C} = \frac{X_C}{R_p}$$

The normal power source for all L, Q, C, and D measurements is the internal, 1-kilohertz oscillator. Balance is indicated by a panel-mounted, zero-centered meter. Meter sensitivity is usually adjustable in order to prevent meter damage when the circuit is far from null and yet give greater resolution at null. Provision is generally made for the connection of external headphones or an oscilloscope as alternative null indicators.

Operation

The typical, manual impedance bridge is a simple instrument to operate and interpret. The component to be measured is connected to the unknown terminals. Then, the appropriate bridge is selected by choosing the R, C_s, C_p, L_s, or L_p mode. Bridge power choices are DC or AC and can be from an internal or external source. With the main power switch turned on, balance can begin.

○ Example of Resistance Measurements

For resistance measurements, internal DC is generally used for bridge power. The *CRL* (capacitance, resistance, and inductance) dial can be placed at midrange, and the range switch can be adjusted until the null indicator comes closest to zero. Then, the *CRL* dial is adjusted until the detector indicates balance. The resistance is the value indicated by the *CRL* dial multiplied by the factor indicated by the range switch.

One more operational feature should be noted. Null detectors are desensitized with internal series multipliers in order to protect them during gross unbalance. As balance is approached, and if fine balancing is required, the detector can be made to be more sensitive with the aid of a sensitivity or multiplier override switch. This momentary-contact switch is engaged to increase detector sensitivity. However, it should be disengaged as soon as balance is indicated so that it can continue protecting the detector.

○ Example of Inductance Measurements

For inductance measurements, the inductor must be connected to the unknown terminals. Internal AC is the normal power source, and L_s or L_p is the mode selected. As noted earlier, two interdependent controls must be used to balance this bridge. The CRL dial is used to balance the reactive characteristics of the component, and the DQ dial balances the resistive effects. As before, the CRL dial must first be placed at midscale, and the best range is selected on the range switch.

Balance is achieved by alternately adjusting the CRL dial and the DQ dial toward null. If null cannot be achieved, it is because the incorrect mode (L_s or L_p) has been selected. Changing modes will allow you to obtain balancing. Once balance is reached, the inductance can be determined by multiplying the CRL dial indication by the range. Quality factor Q is read directly from the DQ dial.

○ Example of Capacitance Measurements

For capacitance measurements, the component is connected to the unknown terminals, the C_s or C_p mode is selected, the internal AC power is applied, and balance is pursued with the CRL and DQ dials. Once again, results are indicated by the CRL dial, the range switch, and the DQ dial. In cases where L and C are being measured, headphones or an oscilloscope may be connected to the external detector terminals for a more definite indication of null.

Specifications

One consideration of an impedance bridge is the range of parameters it can measure. Most measure R, L, C, D, and Q, and some also measure conductance G. Typical ranges for resistance extend from less than 1 ohm to more than 1.1 megohm. Inductance ranges extend from less than 1 microhenry to more than 1100 henrys. Capacitance

ranges spread from less than 1 picofarad to more than 1100 microfarads.

Accuracy is another significant consideration, and it depends
on factors such as the parameter measured, the range used, and the
bridge voltage source. However, an overall accuracy between 0.05%
and 1.0% is typical. This accuracy is usually over a frequency range
from 20 to 20,000 hertz. Other considerations in bridge selection
include portability, battery operation, digital displays, and the options of external sources and detectors.

AUTOMATIC *RLC* BRIDGE

The automatic *RLC* bridge instrument in Figure 7–10 provides a
good indication of the current trends in measurement in general and
in bridges specifically. Integrated circuitry, microprocessors, and
digital displays are becoming more frequently used in such instru-

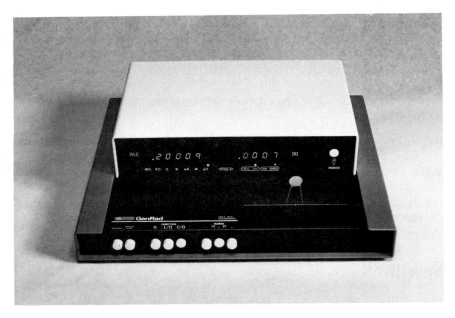

──○ **FIGURE 7–10**
      Automatic *RLC* Tester (Photo courtesy of GenRad, Inc.)

ments. Integrated circuits increase durability, stability, and the overall life of instruments. Also, recalibration is no longer required.

Microprocessor-directed ranging indicates the best range for greatest accuracy and precision. This ranging, along with the five-digit display, provides accuracy to 0.2% or 0.1%. Furthermore, accuracy can be achieved at a rate of three measurements per second.

Internal data processing, such as averaging, reduces operator tasks even further. Internal circuitry for external computer interfacing means that more extensive data monitoring and analysis can be conducted elsewhere. Once again, instruments continue to become faster, more accurate and precise, easier to operate, more versatile, and more compatible with common computer interfacing.

SUMMARY

Bridges are commonly used for the measurement of resistance, inductance, capacitance, quality factor, and dissipation factor. In a bridge, a known magnitude is varied until it equals the magnitude of an unknown component. Equality, or balance, is indicated with the aid of a null detector, usually a galvanometer. An advantage of this process is that the quality of the results is controlled by the quality of the internal standards, *not* the quality of the detector.

The Wheatstone bridge is the basic bridge circuit. It is DC-operated and consists of two series circuits connected in parallel with each other. Voltage is applied at the top and the bottom points, and the galvanometer is connected from one side to the other. The ratio of resistances on one side must be adjusted to match the ratio of resistances on the other side. One side includes the unknown resistance, and the other side includes an adjustable resistor (rheostat) with a calibrated dial. The rheostat is adjusted until the galvanometer indicates the null condition. The unknown value can then be read from the resistor dial.

Inductance can be measured with a Maxwell bridge, which utilizes resistors and capacitors. The capacitors provide the stable and equal, but opposite, phase shift needed for balancing inductance. A resistive branch is also needed in the Maxwell bridge for measuring the series resistance of the inductor. Operation of this bridge is similar to the operation of the Wheatstone, except that the L and Q dials are alternately adjusted until null occurs. Then, inductance

L and quality factor Q values are read from the dials, where $Q = X_L/R_s = R_p/X_L$.

While the Maxwell bridge is used for inductors with a Q that is less than 10, the Hay bridge is used for inductors with a Q between 10 and 100. The Hay bridge balances the parallel resistance of an inductance. The only difference between the two bridges is that the Maxwell bridge has one parallel RC branch, while the Hay bridge has one series branch. Both operate in the same manner. A component of unknown value is connected to two terminals, an internal AC voltage is applied across the bridge, the CRL and DQ dials are alternately adjusted until null occurs, and the L and Q values are read from the two dials.

The Schering bridge is used for the measurement of capacitance and dissipation factor. This bridge also uses an adjustable resistor for balancing the reactance and a parallel RC circuit for achieving the necessary phase shift.

All of the bridge circuits described in this chapter are used as parts of a larger, more versatile instrument called an impedance bridge. Typically, an impedance bridge can measure resistances from less than 1 ohm to more than 1.1 megohm. Inductance ranges extend from less than 1 microhenry to more than 1100 henrys, and capacitance ranges from less than 1 picofarad to more than 1100 microfarads. Accuracy depends on range and parameter, but it can extend from 0.05% to 1.0%.

High-volume, high-quality measurement requirements led to the development of instruments such as the automatic RLC bridge. It includes microprocessors and other integrated circuits as its components, which lead to increased durability and stability, longer life, and the elimination of recalibration. Digital displays and microprocessors provide rapid indication, interpretation, analysis, and storing of measurement data. Typical measurement accuracy is 0.1%.

BIBLIOGRAPHY

Automatic/manual LCR bridge simplifies component measurement, EDN, July 21, 1983.

Automatic bridges speed precision measurements, M. Riezenman, Electronic Design, November 10, 1983.

Passive component bridges meet price-performance demands, Bruce Rohr, Electronic Products, January 11, 1984.

Electronic Instrumentation and Measurements, David A. Bell, Reston Publishing Company, Inc., 1983.

QUESTIONS

1. Name a DC bridge and the parameter it measures.
2. Name three AC bridges and the parameters they measure.
3. Describe the differences between the way an ohmmeter and a DC bridge measures resistance.
4. Define R, L, C, Q, and D.
5. Explain why galvanometer accuracy is less critical in null-balance circuits than it is in open-ended applications.
6. Explain why battery stability is less important in bridges than in open-ended applications.
7. What factors affect the accuracy of a DC bridge? Of an AC bridge?
8. Describe how resistance is measured with a DC bridge.
9. Describe how inductance and quality factor are measured with an AC bridge.
10. Describe how capacitance and dissipation factor are measured with an AC bridge.
11. Name three advantages of an automatic RLC bridge in comparison with a manual bridge system.
12. Name and describe three specifications of a typical impedance bridge.
13. Explain the purpose and operation of the sensitivity switch on an impedance bridge.
14. What effects would a change in oscillator frequency have on the accuracy of an AC bridge?
15. Under what circumstances might a technician apply AC rather than DC when measuring resistance with a Wheatstone bridge?
16. Why does an RLC bridge, such as a Wheatstone bridge, have a potentiometer rather than only fixed resistors, and why does it have a circumference of 15 to 18 inches?
17. What is the relationship between Q and the quality of an inductor? Generally, is a high or a low Q preferred?
18. What is the relationship between D and the quality of a capacitor? Generally, is a high or a low D preferred?

PROBLEMS

1. Calculate the value of the unknown resistance for the Wheatstone bridge of Figure 7–3 if it is balanced when $R_1 = 1$ kilohm, $R_2 = 1$ kilohm, and $R_3 = 275$ ohms.

2. Repeat Problem 1 with R_1 = 10 kilohms, R_2 = 1 kilohm, and R_3 = 750 ohms.

3. Repeat Problem 1 with R_1 = 10 kilohms, R_2 = 3 kilohms, and R_3 = 500 ohms.

4. Repeat Problem 1 with R_1 = 200 ohms, R_2 = 3 kilohms, and R_3 = 500 ohms.

5. Calculate the values of L, R, and Q for the Maxwell bridge of Figure 7-5 if it is balanced when R_1 = 1 megohm, R_2 = 30 kilohms, R_3 = 10 kilohms, and C_1 = 0.01 microfarad.

6. Repeat Problem 5 with R_1 = 1 megohm, R_2 = 50 kilohms, R_3 = 20 kilohms, and C_1 = 0.01 microfarad.

7. Repeat Problem 5 with R_1 = 500 kilohms, R_2 = 40 kilohms, R_3 = 10 kilohms, and C_1 = 0.05 microfarad.

8. Calculate the values of L, R, and Q for the Hay bridge of Figure 7-6 if it is balanced when R_1 = 200 ohms, R_2 = 800 ohms, R_3 = 10 kilohms, C_1 = 0.1 microfarad, and f = 1 kilohertz.

9. Repeat Problem 8 with R_1 = 200 ohms, R_2 = 600 ohms, R_3 = 10 kilohms, C_1 = 0.2 microfarad, and f = 1 kilohertz.

10. Repeat Problem 8 with R_1 = 400 ohms, R_2 = 1 kilohm, R_3 = 10 kilohms, C_1 = 0.2 microfarad, and f = 1 kilohertz.

11. Calculate the values of C, R, and D for the Schering bridge of Figure 7-7 if it is balanced when R_1 = 150 kilohms, R_2 = 10 kilohms, C_1 = 0.001 microfarad, and C_2 = 10 picofarads.

12. Repeat Problem 11 with R_1 = 100 kilohms, R_2 = 12 kilohms, C_2 = 0.001 microfarad, and C_2 = 20 picofarads.

13. Repeat Problem 11 with R_1 = 50 kilohms, R_2 = 8 kilohms, C_1 = 0.05 microfarad, and C_2 = 20 picofarads.

14. Calculate the values of C and R for the capacitance comparison bridge of Figure 7-8 if it balances when R_1 = 10 kilohms, R_2 = 10 kilohms, R_3 = 100 ohms, and C_1 = 0.5 microfarad.

15. Repeat Problem 14 with R_1 = 12 kilohms, R_2 = 8 kilohms, R_3 = 200 ohms, and C_1 = 0.4 microfarad.

16. Repeat Problem 14 with R_1 = 15 kilohms, R_2 = 10 kilohms, R_3 = 50 ohms, and C_1 = 0.4 microfarad.

8

Oscilloscopes

OBJECTIVES

This chapter describes oscilloscope circuits and their basic operation. Capabilities, specifications, and options are also discussed. Finally, oscilloscope operation and typical applications are introduced and described.

    Name and describe seven functional sections of an oscil-
        loscope.
    Name and describe four common specifications of an os-
        cilloscope.
    Describe four options available with some oscilloscopes.
    Describe the basic adjustments that must be made with
        most oscilloscopes.
    Measure voltage, frequency, and phase with an oscillo-
        scope.

THEORY OF OPERATION

There are seven functional sections in a basic oscilloscope, as shown in Figure 8–1. These sections are the cathode ray tube (CRT); the vertical, horizontal, sweep, and trigger circuits; the low-voltage power supply; and the high-voltage power supply. These sections vary in complexity according to the capabilities of the oscilloscope. Nevertheless, they share some common features, which are described in the subsections that follow.

151

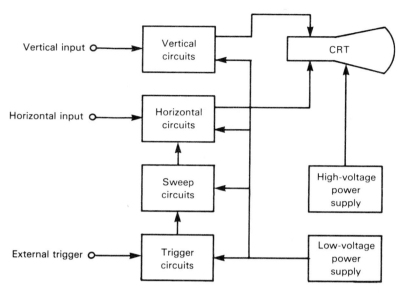

FIGURE 8–1
Block Diagram of Basic Oscilloscope

Cathode Ray Tube

The primary function of all circuits in an oscilloscope is to establish and maintain a trace (or traces) on the face of a *cathode ray tube* (CRT). As Figure 8–2 shows, a CRT has a heater, a cathode, a grid, anodes, and plates, just like other electron tubes. It is, however, much more complicated.

The current through the heater raises its temperature, which, in turn, heats the cathode to a temperature near 1650 degrees Celsius. At this temperature, electrons boil from the cathode through a process called thermionic emission, and they form a negative space charge in the surrounding area. With the cathode voltage near −1600 volts and the first anode at ground, a difference of 1600 volts exists, causing the free electrons to accelerate toward the first anode. Since this preacceleration anode is a cylinder, most electrons speed right on through it, although the electron beam does tend to widen. A small, negative (adjustable) voltage is then applied to the focusing anode, and it narrows the beam.

The second, or acceleration, anode is at the same voltage as the first anode. It reinforces the effect of the first anode. The electrons continue on and collide with the screen, which is at a potential of up to +15 kilovolts or so. The phosphor-coated screen glows at

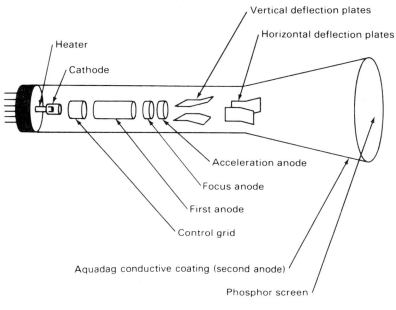

Vertical deflection plates

Horizontal deflection plates

Heater

Cathode

Acceleration anode

Focus anode

First anode

Control grid

Aquadag conductive coating (second anode)

Phosphor screen

──o **FIGURE 8–2**
Cathode Ray Tube

the point of impact. The electrons then flow along the conductive coatings of the screen and the *aquadag,* to the high-voltage power supply, and back to the cathode to replace the electrons emitted.

The intensity of electron flow is controlled with voltages applied to the control grid. A negative grid voltage repels the electrons back toward the cathode, while a reduction of this voltage allows the flow to increase. This process controls screen brightness. Voltage applied to the focus anode concentrates the electron flow into a narrow beam, causing a sharply defined spot on the screen. The intensity and focus controls adjust the voltages applied to these two anodes.

Figure 8–3 shows how the application of voltages to the vertical deflection plates redirects the beam as it passes through. Negative bias on the top and positive bias on the bottom move the beam downward. The amount of deflection is directly proportional to the magnitude of the voltage, and the direction depends on polarity. The *deflection sensitivity* of a CRT describes the ratio of deflection plate voltage V_p to beam deflection D on the screen:

$$\frac{V_p}{D} = \frac{2 \times V_a \times d}{L \times \ell}$$

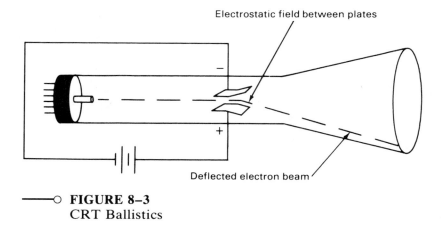

Electrostatic field between plates

Deflected electron beam

──○ **FIGURE 8–3**
CRT Ballistics

where V_a = voltage of second anode, d = distance between deflection plates, L = length of beam from center of plates to screen, and ℓ = length of plates.

○ **Example of Deflection Sensitivity**

Consider a CRT with a second-anode voltage of 600 volts, 2.5-centimeter plates that are 1 centimeter apart, and a screen that is 16 centimeters beyond the center of the plates. The deflection sensitivity of this CRT is determined as follows:

$$\frac{V_p}{D} = \frac{2V_a d}{L\ell} = \frac{2 \times 600 \times 1}{16 \times 2.5} = 30 \text{ V/cm}$$

A deflection sensitivity of 30 volts per centimeter can also be considered in this way. If this scope had an 8-centimeter screen, it would require vertical deflection voltages of +120 and −120 volts.

Figure 8–4A depicts the case where no voltage is applied. But just as voltages on the vertical deflection plates move the beam up or down, voltages on the horizontal plates redirect the beam left or right. For example, a negative voltage on the right horizontal plate and a positive voltage on the left plate moves the beam toward the left, as shown in Figure 8–4B. Figure 8–4C shows that a positive voltage on the top vertical plate and a negative voltage on the bottom plate moves the beam up.

Combining both the forces shown in Figure 8–4B and 8–4C will move the beam up and to the left, as shown in Figure 8–4D. Since the horizontal and vertical plates are electrically independent, dif-

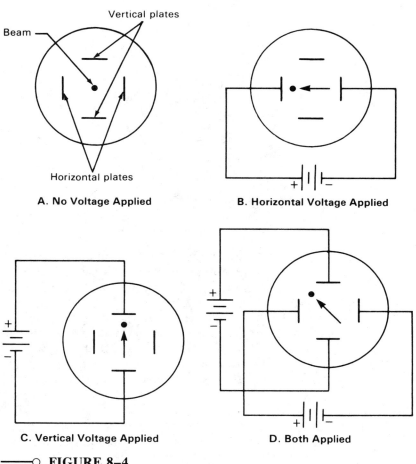

Horizontal plates

A. No Voltage Applied   **B. Horizontal Voltage Applied**

C. Vertical Voltage Applied   **D. Both Applied**

──O **FIGURE 8–4**
Horizontal and Vertical Deflection in CRT

ferent voltages can be applied to each set. The voltages for these plates come from the horizontal amplifier and the vertical amplifier.

Two points should be noted about beam displacement. First, the magnitude of beam displacement is linear; that is, it is directly proportional to the magnitude of the vertical-input voltage. Second, since only an electron beam is moved, the change is made quickly and requires little energy.

Quite often, an oscilloscope is used to view more than one signal at the same time. In this situation, we can use a *dual-beam CRT,* such as the one shown in Figure 8–5. This tube has two independent beams, each controlled by separate circuits. Each has its

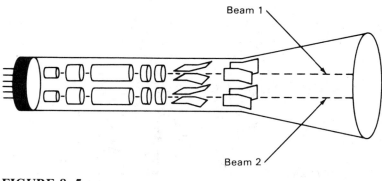

───○ **FIGURE 8–5**
Dual-Beam CRT

own cathode, anodes, and deflection plates. A more common method for obtaining multiple traces, though, is to rapidly switch a single beam between two or more sources. This method will be discussed in detail shortly.

Another form of the CRT is used in *storage oscilloscopes,* instruments with the ability to maintain a screen image even after the applied signal is removed. During normal operation, the storage oscilloscope operates like any other scope. When an image is to be stored, though, the operation changes. Electrons leave the *write gun,* shown in Figure 8–6, at a very high energy level. When they reach the storage mesh, they dislodge electrons at the point of impact. The dielectric material of the storage mesh now becomes positively charged. This charge remains even when the beam is discontinued. Thus, the image is stored.

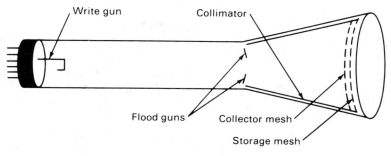

───○ **FIGURE 8–6**
Storage CRT

For viewing of the image, low-velocity electrons are emitted from the *flood guns* and formed into a cloud by the *collimator*. The collector mesh gathers many electrons and becomes negatively charged, except in the area aligned with the image on the storage mesh. Remember that part of the storage mesh is positively charged. At this point, flood gun electrons are repelled back to the CRT, except for those electrons that penetrate the image section of the storage mesh. A screen potential near 20,000 volts accelerates these electrons, causing a visible image in that area. The image can be erased by discharging the storage mesh and allowing the flood gun electrons to flow on toward the screen.

Vertical Circuits

A signal to be measured is applied to a vertical input of an oscilloscope. It goes through an AC–Gnd–DC switch, where a capacitor is placed in series (AC) or is bypassed (DC), or the input to the scope's amplifer is disconnected from the external jack and is grounded. In the AC mode, the capacitor allows small AC signals to be viewed without the overriding influence of the large DC combined with it. From here, the signal proceeds to an attenuator and a preamplifier, where it is decreased or increased in amplitude, as needed for adequate deflection.

The block diagram in Figure 8–7 represents the vertical circuits of a *dual-trace oscilloscope*. While this scope does not have dual beams, it does have the capability of producing two images, one from input A and one from input B. It produces its two images by rapidly switching between the two inputs with the aid of an electronic switch. The vertical mode switch provides a choice among input A, input B, chopped, or alternate. With an input A or an input B choice, the single selected source can be viewed. The chopped and alternate modes allow two signals to be viewed at once. The switching is controlled by the sweep circuits and will be described later when sweep circuits are discussed.

The signal proceeds through a delay line, which delays the signal 100 to 200 nanoseconds before it reaches the vertical deflection amplifier. The delay is introduced in order to ensure that the signal arrives at the CRT after the horizontal signals arrive. The horizontal signals have an inherent delay of their own. An *LC* current is used to introduce the vertical signal delay.

A direct-coupled driver followed by a double-ended, vertical output amplifier brings the input to the two vertical deflection plates as differential signals. The attenuator-amplifier circuits are con-

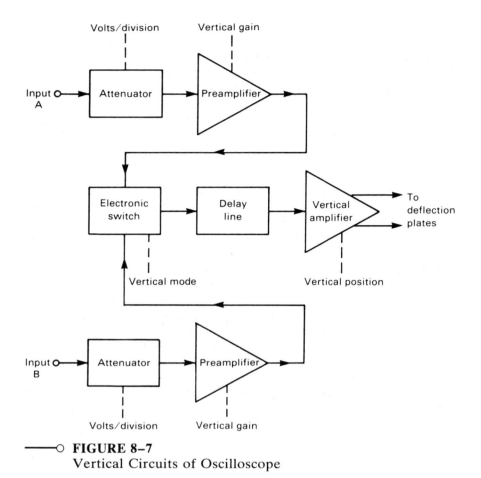

——O **FIGURE 8–7**
Vertical Circuits of Oscilloscope

trolled by vertical range or gain controls, which are generally designated in volts per centimeter (V/cm) or millivolts per centimeter
(mV/cm).

The vertical position control is usually part of the final stage.
It is a potentiometer that allows the operator to move the display
up or down for alignment with the CRT grid during measurement.

Horizontal Circuits

As shown in Figure 8–8, the horizontal circuits of most oscilloscopes
can respond to two signal sources, internal and external. Operator
selection is made with a switch labeled time and X–Y. In the X–Y

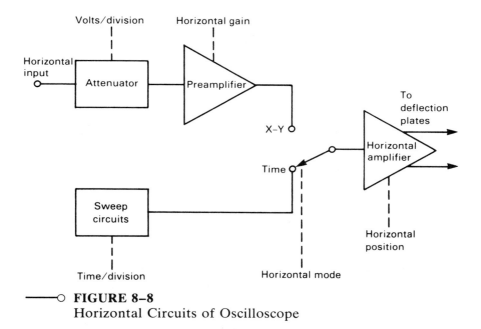

———O **FIGURE 8–8**
Horizontal Circuits of Oscilloscope

mode, signals connected to the external horizontal input are processed in a manner similar to the way they are processed in the vertical circuit; that is, they are attenuated or amplified. The horizontal path, though, usually has less amplification and fewer ranges available. Once again, the signal ends at two deflection plates (horizontal) as differential voltages from a double-ended amplifier.

When an external signal is applied to the horizontal input, it can be compared with another signal that is applied to the vertical input. Most of the time, however, oscilloscopes are not used to evaluate one signal (vertical) as a function of another (horizontal). Instead, oscilloscopes are used to measure some signal as a function of time. In the time mode, a signal is produced by the internal sweep circuitry and is applied to the horizontal amplifiers.

Sweep Circuits

The typical *sweep circuit,* represented in Figure 8–9, begins with a ramp generator, which produces a sawtooth-shaped signal. As indicated in Figure 8–9, many oscilloscopes have an additional input capability called the Z, or intensity, input, and it is often located on the rear of the cabinet. It is connected to the blanking circuit, and

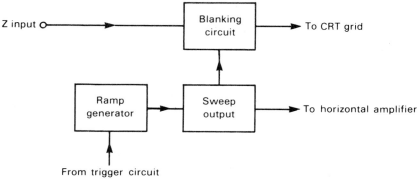

—————○ **FIGURE 8–9**
Sweep Circuits of Oscilloscope

it allows the operator to simultaneously view the interrelationship between three signals. One signal connected to the Y input moves the beam up or down; a second signal connected to the X input moves the beam right or left; and a third signal connected to the Z input makes the display darker or lighter (it is similar to the brightness control on a black and white TV).

A typical sawtooth signal produced by the sweep circuit is shown in Figure 8–10A. The signal begins at a negative point and proceeds up through zero and on to an equal positive point in a linear manner.

A common ramp generator circuit consists of an oscillator energized from a constant-current source. Fixed ranges are selected by switching in different values of capacitance. Sweep rates in between are achieved by adjusting a potentiometer in the current source network.

When the sweep signal is applied to the horizontal deflection plates, the beam moves from left to right at a constant rate, rapidly back to the left side, and then repeats the cycle. The rate at which this sweep occurs, called the *sawtooth frequency*, is selected by the time/division control of the scope (see Figure 8–8).

The sharp drop that causes a retrace is detected and sent on to the blanking circuit. This signal, in turn, commands the control grid to turn the beam off momentarily during retrace to eliminate a useless and otherwise distracting line on the screen. This procedure is called *retrace blanking.*

As mentioned earlier, the period of the sweep sawtooth waveform is controlled by the time/division switch. Normally, we select a horizontal sweep time equal to the period of the vertical signal being viewed. In this way, one cycle of the vertical signal will be

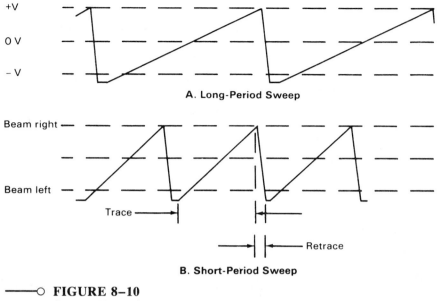

A. Long-Period Sweep

B. Short-Period Sweep

——○ **FIGURE 8–10**
Sweep Waveforms

displayed (Figure 8–10A). When a higher-frequency signal is to be displayed, a short-period sweep (Figure 8–10B) is used.

In the section on vertical circuits, we noted that two methods are commonly used for viewing two signals at once. These methods are alternate and chopped sweep. In the *alternate sweep* mode, the electronic switch in the vertical circuit sends input A on to the vertical deflection amplifier. After one sweep of the horizontal deflection circuit, the electronic switch changes and sends input B to be displayed. The switch toggles back and forth at the command of the sweep generator, alternating the display of the two inputs. This method works well for high-frequency signals (high sweep rates).

In the *chopped sweep* mode, a free-running multivibrator rapidly switches back and forth, alternately sending segments of one input and then the other on to the vertical amplifiers. The chopped sweep procedure is illustrated in Figure 8–11. The switching rate is usually around 500 megahertz, and, ideally, it is not a multiple of the waveform frequencies. While segments of each waveform are left out on a given pass, they are usually filled on another pass. This method works well for low-frequency signals. At high frequencies, the gaps will probably be seen.

As shown in Figure 8–12, *signal sampling* is a method by which a signal with a frequency beyond the sweep range of a scope can be

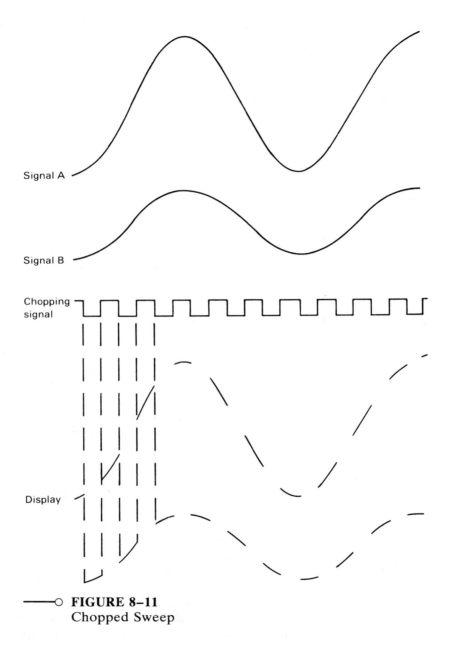

─────○ **FIGURE 8–11**
Chopped Sweep

viewed. With this system, samples of the input signal are taken, stored, and then displayed on the CRT as a function of a sweep rate within the range of the scope. The result is an approximation of the original signal but at a lower frequency.

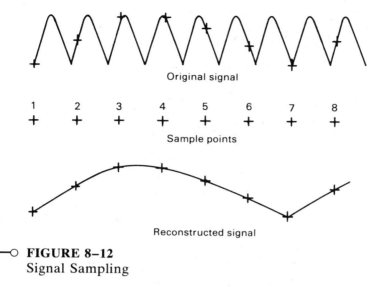

Original signal

1   2   3   4   5   6   7   8

Sample points

Reconstructed signal

───○ **FIGURE 8–12**
Signal Sampling

Trigger Circuits

The block diagram for a *trigger circuit* is shown in Figure 8–13. The trigger circuit operates in conjunction with the sweep circuit and controls the time when a sweep retrace occurs. By doing so, it controls the synchronization between the sweep signal and the signal to be viewed. Synchronization, in turn, ensures that the beam continually sweeps the same path, making one continuous trace on the CRT face.

The typical input signal to an oscilloscope is a sine wave that continually rises and falls. Likewise, the ramp-shaped sweep voltage also rises and falls. Ideally, their periods should be almost identical. The role of the trigger circuitry is to ensure that the horizontal sweep begins (point 1 in Figure 8–14) at a specific time in the period of the vertical input signal. This time is usually the time when the input signal crosses zero in the positive direction, as shown in Figure 8–14. Without this synchronization, there will be no controlled relationship between the vertical input and the sweep signals, and the display will not remain stable.

With the trigger source switch in the INT (internal) position, a sample of the vertical signal is amplified and sent on to a Schmitt trigger. This circuit monitors the vertical signal and detects the point when it reaches a selected level (trigger level) while moving toward a selected polarity (trigger slope).

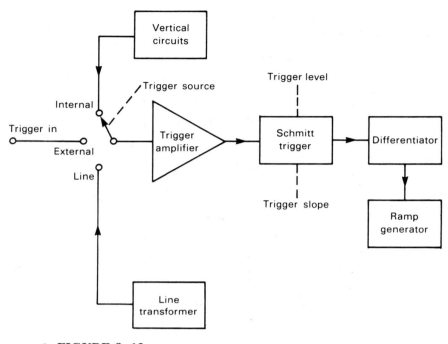

——○ **FIGURE 8–13**
Trigger Circuit of Oscilloscope

When the trigger level is reached, a square wave is sent to a differentiator, where it becomes a pulse. This pulse is sent to the ramp generator and is the command for the sweep to begin the trace. The trigger level is position 5 in Figure 8–14. The time loss between the command to trace and the restart of the horizontal sweep is one of the losses compensated for by the vertical delay line.

Other options for trigger sources are line and external. The line mode allows the operator to view 60-hertz signals. The external mode is used when an external trigger command is desired, such as when the operator is viewing a digital circuit and synchronizing with its internal clock.

Figure 8–15 shows the effects of the trigger slope and trigger level controls. In Figure 8–15A, a + slope causes the sweep to begin when the trace is on the way up. The − 2 level causes it to begin at a point below zero. A + 5 level causes the sweep to begin at a point not only further from zero but also above it, as shown in Figure 8–15B. Negative (−) slope triggering is shown in Figure 8–15C.

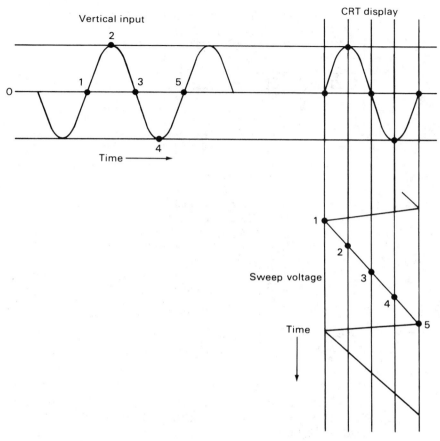

─────○ **FIGURE 8–14**
Display Synchronization

Power Supplies

There are two types of power supplies in most oscilloscopes, low-voltage and high-voltage, as diagrammed in Figure 8–16. The low-voltage supplies operate in the hundreds of volts (or less) and power the amplifier, sweep, calibrator, and bias circuits. Both positive and negative voltages are needed for the differential amplifiers. The calibrator needs an accurate, stable voltage.

Excellent filtering of the power supplies is essential because of the high amplifier gain involved. Regulation is essential because the amplifiers are direct-coupled.

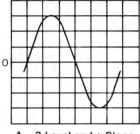
A. -2 Level and + Slope

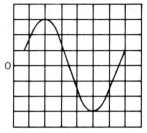
B. +5 Level and + Slope

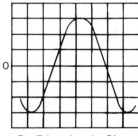
C. -7 Level and - Slope

——○ FIGURE 8–15
Trigger Levels and Trigger Slopes

The high-voltage supplies operate in the thousands of volts and provide CRT acceleration voltages. For most oscilloscopes, DC voltages from 1600 to 2200 volts are typical, although 20,000 volts are not uncommon in storage oscilloscopes.

SPECIFICATIONS AND OPTIONS

The cost of an oscilloscope, such as the one pictured in Figure 8–17, can range from hundreds to thousands of dollars, depending on the capabilities required and the options desired. Prices increase with increased refinements in bandwidth and sensitivity and with added features. Generally, an oscilloscope is selected in the same way other instruments are selected. That is, requirements are identified, prioritized, and then balanced against available funds. Some typical specifications and options for oscilloscopes are discussed in this section.

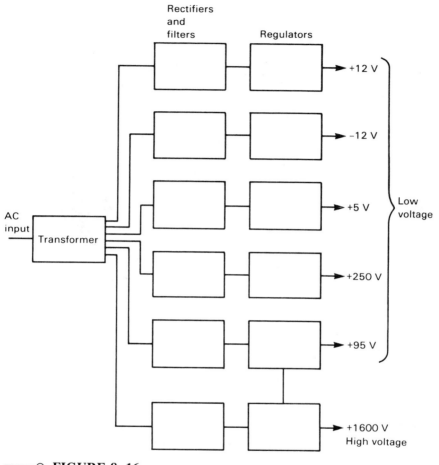

———○ **FIGURE 8–16**
Power Supplies of Oscilloscope

Specifications

The specifications of an oscilloscope describe important features and allow comparison of that instrument with other brands and models. The importance of each specification varies with the application. Typical specifications and their definitions are listed here in order of general importance:

   1. The *bandwidth,* which is the frequency range between the upper and lower limits, is the upper frequency limit for the

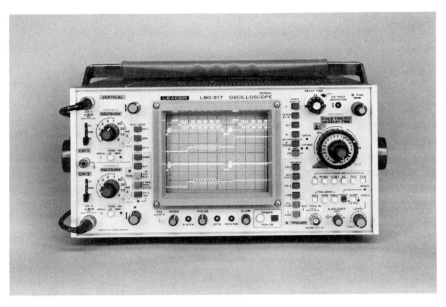

—○ **FIGURE 8–17**
Quad-Trace, Dual–Time Base Oscilloscope (Photo courtesy
of Leader Instruments Corporation)

oscilloscope, since most instruments operate down to DC.
A bandwidth of 50 megahertz is typical, and over 1 gigahertz
is available.

2. The *deflection factor* is the smallest voltage measurable with
   the instrument. Like bandwidth, this specification has a direct relationship with cost. A factor of 1 millivolt per centimeter (mV/cm) is typical, and 1 microvolt/centimeter (μV/
   cm) is available.

3. The *number of channels* is the number of independent vertical inputs available—that is, the number of signals that can
   be viewed simultaneously. One channel is typical, two channels are quite common, and four channels are available.

4. The options available for the cabinet include standard bench-
   top, portable (small and rugged), mainframe (allows for plug-
   ins), and rack-mounted.

5. *Persistence* is the duration of image retention and is controlled by the phosphor coating on the CRT. A long- or
   short-persistence CRT is selected according to the needs of
   the intended application. Some oscilloscopes have a control
   providing variable persistence.

Displays

The typical oscilloscope has a 5-inch rectangular screen covered with a calibrated *graticule,* which is a grid of horizontal and veritcal lines 1 centimeter apart. Many oscilloscope displays have screen illumination lamps along the edge of the graticule, which make the grid lines more visible. Beam finders indicate the direction in which a missing trace has gone off the screen.

Light filters, viewing hoods, and camera mounts can be attached for better viewing or for photographing the trace. As noted earlier, storage oscilloscopes allow an image to be retained and redisplayed up to hours later. Digital displays using LEDs or LCDs present precise indications of voltage and/or time. This feature eliminates one of the most serious oscilloscope limitations—user interpretation error caused by lack of precision in the display.

Vertical Options

There are many vertical options available, both built-in and plug-in. The model shown in Figure 8–18 has the capability of accepting two vertical plug-ins and two horizontal plug-ins. The plug-in on the left end is a dual-trace amplifier, which allows a user to view two signals at once. It often features alternate or chopped sweep, which selects the manner in which multiple channels are viewed. In the alternate sweep mode, one sweep of one signal is made, followed by one sweep of the second signal, followed by one sweep of the first signal, and so on. The chopped sweep mode provides a bit of one, a bit of a second, a bit of the first, and so on, until a complete image of each signal has been produced.

Another plug-in is a wide-band amplifier (DC to 600 megahertz). Finally, one other vertical option available is a differential amplifier. This high-gain (10 microvolts per centimeter) amplifier allows the comparison of two signals by displaying the difference between them.

Horizontal Options

The plug-in at the far right in Figure 8–18 is a horizontal, dual–time base unit. With this option and multiple vertical inputs, two signals of different frequencies can be viewed at once. For a scope without dual–time base capability, two signals must have the same frequency or be harmonics of each other in order to be viewed simultaneously.

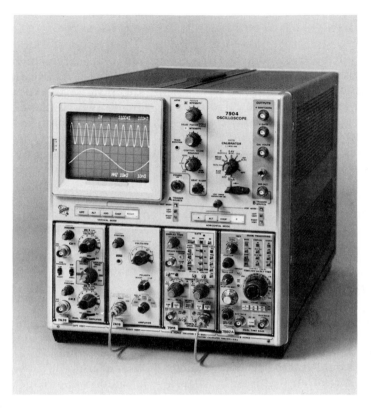

○ **FIGURE 8–18**
Modular, General-Purpose Oscilloscope (Photo courtesy of
Tektronix, Inc.)

The second plug-in unit from the right end is a timer/counter, which
provides direct, accurate frequency measurement. Delayed time base
allows a small portion of a signal to be selected, intensified, and
expanded to full-screen width for more accurate measurement or for
better viewing.

Probes

The high-impedance, or divider, probe in Figure 8–19 is typical of
the probes most commonly used. Its internal resistance increases the
apparent input impedance of an oscilloscope from the typical 1 megohm to 10 megohms. However, it also attenuates the incoming signal
by a factor of 10 at the same time, so a 25-millivolt signal appears
to the vertical circuitry as 2.5 millivolts. As a reminder, probes are

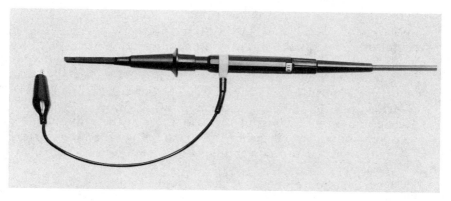

—○ **FIGURE 8–19**
Oscilloscope Probe (Photo courtesy of Leader Instruments Corporation)

generally marked ×1 or ×10 to indicate the impedance and voltage multiplier factor.

Multiplier ratio, capacitance, and method of attachment are just three of the variables to consider in probe selection. For example, current probes allow measurement without circuit interruption by producing a voltage proportional to the field strength around a conductor. High-voltage probes are used to divide large voltages, protecting both operator and instrument.

APPLICATIONS

Voltage, period, and phase are the most common measurements made with an oscilloscope. None are difficult to do, although the number of controls involved increases the likelihood of error. Some controls *should be* adjusted in a particular way for best results. Other controls *must be* adjusted in a particular way for correct results. Our discussion begins with controls and their adjustment.

Controls and Adjustments

The names and functions of common oscilloscope controls are listed here. The abbreviations for these controls that are printed on the instrument will vary with brand and model. As always, check the manual if you are uncertain about the name or function of a control on the scope you are using.

On-off is the main power control.

Focus controls the sharpness of the trace.

Intensity controls the brightness of the trace.

AC–Gnd–DC switches a capacitor in or out, or grounds the amplifier input, according to the signal to be viewed.

Horizontal position moves the trace left or right.

Vertical position moves the trace up or down.

Scale illumination controls the brightness of the grid lines.

Trigger source selects either the signal being viewed, an external signal, or the power line as the sweep-triggering command.

Trigger level determines what the level of the reference signal should be when triggering is to occur.

Trigger slope determines whether triggering will occur on the positive or the negative slope of the reference signal.

Vertical range, or volts per division, selects the vertical calibration range.

Vertical gain, or vernier, is used to uncalibrate the vertical scale for more convenient viewing.

Horizontal range, or time per division, selects the horizontal calibration range.

Horizontal gain, or vernier, is used to uncalibrate the horizontal scale for more convenient viewing.

A properly adjusted focus control produces a sharply defined trace. The intensity should be adjusted so that the waveform is just bright enough to see clearly. Since CRT age is reduced by excessive brightness, *never leave a spot on the screen.*

Before any measurements are made, probe compensation is required. As Figure 8–20 shows, a scope probe has internal resistance (R_1) and an adjustable capacitance (C_a) within the probe. This capacitance is varied, often by rotating the probe handle, to produce

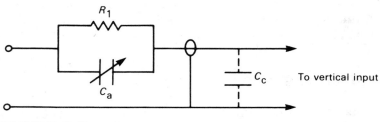

FIGURE 8–20
Probe Circuit

a value that will compensate for the capacitance of the cable (C_c) and of the scope input.

Proper compensation is achieved when a square-wave input signal produces sharp corners, as shown in Figure 8–21A. Figure 8–21B shows that undercompensation causes undershooting. Figure 8–21C shows that overcompensation causes overshooting. With many oscilloscopes, a square wave for probe compensation is available on the front panel.

Voltage Measurement

The first step in measuring a voltage is to check for *calibration*—that is, to check that the scope's output is as specified by the manufacturer. One of two methods is usually used. In one method, the

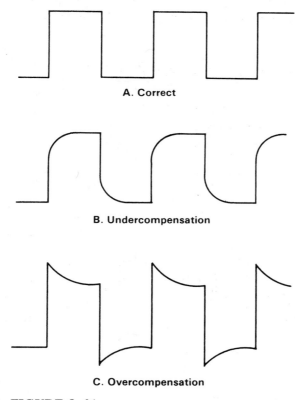

A. Correct

B. Undercompensation

C. Overcompensation

─○ **FIGURE 8–21**
Probe Compensation

vertical vernier or gain control is placed in the calibrated position. This position is at the end of the rotation and is indicated by a switch detent (or catch). In the second method, the vertical range switch is placed in the calibrate position, which produces a square wave on the screen. Calibration is then achieved by adjusting the vernier control until the square wave reaches the specified size, usually 4 or 6 centimeters peak-to-peak.

After calibration, the instrument should be zeroed. Zeroing is done by switching the input to ground or by grounding the probe and adjusting the vertical position for a straight line centered vertically on the screen. A zeroed display is shown in Figure 8–22A.

With zeroing completed, you can connect the unknown voltage to vertical input, select AC or DC, and select the range that produces the largest, completely visible display. Remembering that a positive voltage makes the beam go up and a negative voltage makes it go down, you can determine the voltage by multiplying the number of divisions of displacement by the selected range. The following example shows some calculations.

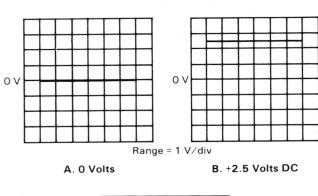

Range = 1 V/div

A. 0 Volts   B. +2.5 Volts DC

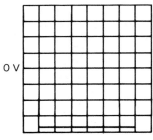

C. -3.8 Volts DC

——○ **FIGURE 8–22**
Voltage Measurement Displays

○ **Example of Calculating DC Voltage from Oscilloscope Trace**

For the trace shown in Figure 8–22B, the voltage is calculated as follows:

$$+2.5 \text{ cm} \times 1 \text{ V/cm} = +2.5 \text{ V}$$

For the trace shown in Figure 8–22C, the voltage is calculated as follows:

$$-3.8 \text{ cm} \times 1 \text{ V/cm} = -3.8 \text{ V}$$

The trigger controls must also be adjusted in order to get an acceptable waveform. For DC measurement, the automatic or line positions are satisfactory. With these controls, the trace repeats itself automatically on its return or whenever the line voltage repeats its cycle. While this selection also works for viewing line frequency signals, it is unsatisfactory with any others. Since the trigger circuit controls the exact point at which the internally generated sweep signal sends the beam across the screen, it must not be too high. If it is, the signal may never reach that high point, and the sweep will never begin.

For AC voltage measurement, you select the vertical range that produces the highest display without going beyond the screen. Count the number of divisions of peak-to-peak deflection, and multiply that number by the range. The rms value of this voltage is determined by multiplying the peak value (peak-to-peak divided by 2) by 0.707.

○ **Example of Calculating AC Voltage from Oscilloscope Trace**

The value of the voltage in Figure 8–23A is determined as follows:

$$4 \text{ div} \times 50 \text{ mV/div} = 200 \text{ mV pp}$$

$$V_{rms} = \frac{200 \text{ mV}}{2} \times 0.707 = 70.7 \text{ mV rms}$$

The voltage in Figure 8–23B is calculated as follows:

$$6 \text{ div} \times 2 \text{ V/div} = 12 \text{ V pp}$$

$$V_{rms} = \frac{12 \text{ V}}{2} \times 0.707 = 4.2 \text{ V rms}$$

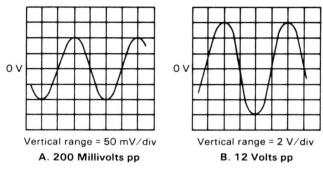

Vertical range = 50 mV/div   Vertical range = 2 V/div
A. 200 Millivolts pp   **B. 12 Volts pp**

——○ **FIGURE 8–23**
AC Voltage Measurement

The external trigger position is used whenever one-shot (single-sweep) events are to be viewed or when synchronization with an external reference is desired. Then, the voltage applied to the circuit being evaluated is connected also to the external trigger input terminal and used as the trigger command. Sweep begins when the test circuit input is applied to the test circuit. What you observe on the CRT is the time it takes the test circuit to respond and what the circuit's response is.

Frequency Measurement

The frequency of a signal is determined by measuring the time elapsed for one cycle and calculating its reciprocal. You begin by connecting that signal to the vertical input, adjusting the vertical range for a convenient height, and selecting the lowest horizontal range that shows one complete cycle. The horizontal vernier control must be in the calibrated position for accurate measurement.

After these adjustments are made, the waveform should be vertically centered and horizontally positioned toward the left so that the zero crossing intersects a grid intersection. This positioning is shown in Figure 8–24. The period or time for one cycle is then determined by counting the number of horizontal divisions between the two zero crossings and multiplying that number by the horizontal range.

○ **Example of Calculating Frequency from Oscilloscope Trace**

The frequency *f* of the waveform in Figure 8–24A is calculated in the following way:

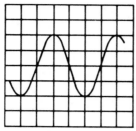 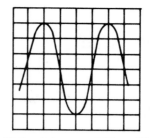

Horizontal range = 20 μs/div Horizontal range = 100 ms/div

A. 80 microseconds per cycle = **B. 400 milliseconds per cycle =**
 12.5 kilohertz   **2.5 hertz**

──○ **FIGURE 8–24**
Frequency Measurement

period $= T = 4$ div $\times$ 20 μs/div $= 80$ μs/cycle

$$f = \frac{1}{T} = \frac{1}{0.000080 \text{ s/cycle}} = 12{,}500 \text{ Hz} = 12.5 \text{ kHz}$$

The frequency of the waveform in Figure 8–24B is determined as follows:

$$T = 4 \text{ div} \times 100 \text{ ms/div} = 400 \text{ ms/cycle}$$

$$f = \frac{1}{T} = \frac{1}{0.400 \text{ s/cycle}} = 2.5 \text{ Hz}$$

Another method of measuring frequency is to compare the unknown frequency with a known frequency. That is, the unknown frequency is connected to the vertical input, and a source of variable and known frequency is connected to the horizontal input. The scope's horizontal input selector must be on external horizontal or its equivalent rather than on a time-per-centimeter range. The known frequency is then varied until a circle appears on the screen, as shown in Figure 8–25A. This trace is called a *Lissajous pattern,* which is the name given to any trace produced by combining two sine wave signals, one from the horizontal channel and one from the vertical channel.

The circle of Figure 8–25A is developed because both the horizontal and the vertical circuits are commanding one cycle (up-down or left-right) at exactly the same rate. A circle indicates a 1:1 ratio between the two signals. An important feature of this method is that the scope has little to do with the quality of the measurement. Ac-

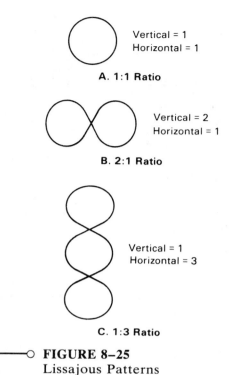

Vertical = 1
Horizontal = 1

A. 1:1 Ratio

Vertical = 2
Horizontal = 1

B. 2:1 Ratio

Vertical = 1
Horizontal = 3

C. 1:3 Ratio

○ **FIGURE 8–25**
Lissajous Patterns

curacy is totally dependent on the accuracy of the reference generator.

Other ratios can also be used if the reference generator's range excludes the unknown frequency. Figure 8–25B shows what occurs when the vertical signal's frequency is twice the horizontal frequency. The ratio is 2:1, with the vertical frequency twice the horizontal frequency.

○ Example of Calculating Frequency from Lissajous Pattern

If the frequency of the horizontal signal in Figure 8–25B is 400 hertz, the vertical frequency can be determined as follows:

$$\frac{\text{vertical}}{\text{horizontal}} = \frac{2}{1}$$

$$\frac{\text{vertical}}{400\,\text{Hz}} = \frac{2}{1}$$

$$\text{vertical} = 400\,\text{Hz} \times 2 = 800\,\text{Hz}$$

A good way to remember which signal is 2 and which is 1 is to look at the waveform and answer the following questions: How many times does it touch the top? It touches the top 2 times. How many times does it touch a side? It touches a side 1 time. What makes it touch the top? The vertical signal does. What makes it touch the side? The horizontal signal does. Hence, the vertical signal is twice the horizontal signal.

Figure 8-25C shows a 1:3 ratio, with the vertical frequency one-third of the horizontal frequency. If the horizontal frequency is 6 kilohertz, the vertical frequency is as follows:

$$\frac{vertical}{horizontal} = \frac{1}{3}$$

$$\frac{vertical}{6000\,Hz} = \frac{1}{3}$$

$$vertical = \frac{6000\,Hz}{3} = 2000\,Hz$$

Pulse Measurement

When you are measuring pulses, it is generally desirable to use the pulse source as an external trigger. Figure 8-26 shows the essential parameters of a pulse. They are amplitude, rise time, fall time, pulse width, and repetition rate (frequency). The *rise time* is the time it takes a pulse to rise from 10% to 90% of its amplitude. The *fall time* is the time it takes a pulse to fall from 90% to 10% of its amplitude. *Pulse width* is the time that elapses between the 50% level points. Repetition rate, or frequency, is the number of pulses per second.

Phase Measurement

An oscilloscope can be used to determine the *phase relationship* between two voltages of the same frequency, that is, the number of degrees difference between the points where each signal crosses 0 volts on the way up. One voltage is connected to the vertical input in the normal manner. The other voltage is connected to the horizontal input, with the horizontal range control placed on a voltage rather than a time position. With these connections and the trigger on automatic, an elliptical waveform similar to the one in Figure 8-27A should appear. This waveform results from an X-Y scope connection.

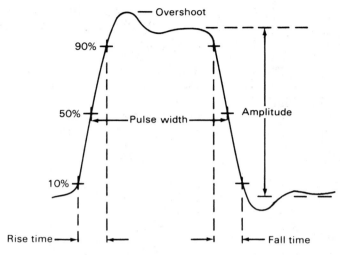

───○ **FIGURE 8–26**
Pulse Measurement

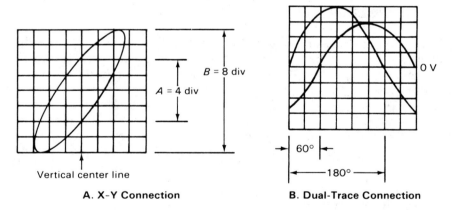

A. X-Y Connection   B. Dual-Trace Connection

───○ **FIGURE 8–27**
Phase Measurement

○ **Example of Calculating Phase from Oscilloscope Trace**

Phase is determined by dividing the number of divisions between the intersections of the waveform with the vertical center line
(A in Figure 8–27A) by the total height of the waveform (B in the
figure). The quotient is the sine of the phase angle. In this case, the
distance between the intersections (A) is 4 divisions, and the total

height (B) is 8 divisions. The sine, therefore, is 4/8, or 0.5, and the phase angle is 30 degrees.

Figure 8–27B shows another way to measure the phase relationship between two signals. One signal is connected to each input of a dual-trace scope. Both inputs must be zeroed vertically across the center line. Then, the horizontal gain should be uncalibrated so that one waveform has a convenient width, such as 180 degrees in 6 divisions. Thus, each horizontal division represents 30 degrees of phase shift. Since the left waveform crosses zero on the way up, 2 divisions before the right waveform, the left waveform *leads* the right waveform by 60 degrees. Thus, in this method, we can determine whether the phase angle is positive or negative.

SUMMARY

There are seven functional sections of an oscilloscope: cathode ray tube, vertical amplifier, horizontal amplifier, low-voltage power supply, high-voltage power supply, and sweep and trigger circuits. The primary component is the cathode ray tube. All other circuits support it. A CRT operates by emitting electrons from a cathode and accelerating them toward a screen, which illuminates at the point of impact. Along the way, this beam of electrons is focused, accelerated, and deflected by various electrostatic fields.

Signals to be measured are connected to the vertical input. Through a combination of amplifiers and attenuators, signal amplitude is increased or decreased, as needed, to produce a measurable image. Double-ended output amplifiers provide differential voltages for the vertical deflection plates. A positive voltage on the upper plate and a negative voltage on the lower plate moves the beam upward. Reversing the polarity moves the beam downward. Beam deflection is proportional to input voltage. The horizontal circuits operate in a similar manner, moving the beam left or right according to the polarity of the applied signal.

During normal operation, the signal to be measured is applied to the vertical input connector. Since this signal can only move the beam up or down, an internal signal is generated in the sweep circuit and applied to the horizontal plates. This sawtooth wave sweeps the beam from left to right and then returns it rapidly for another trace. Sweep rate is controlled by the time/division control and is adjusted to coincide with the period of the input signal.

Oscilloscope power supplies provide a range of low and high DC voltages. Differential amplifiers require positive and negative voltages, usually under 100 volts. Bias and calibration circuits also require smooth, stable DC. Acceleration voltage requirements range from 2000 to 20,000 volts, the latter voltage being required for storage scopes.

Oscilloscope specifications describe important characteristics of an instrument and allow comparison of one scope with others. Typical specifications describe bandwidth, deflection factor, number of channels, cabinet types, and persistence. Their order of importance depends on the user's needs.

A wide range of options is available for most oscilloscopes. The options can be built-in, plug-in, or add-on options. Add-on options include light filters, viewing hoods, and camera mounts. Built-in options include dual vertical channels, dual time base, and delayed sweep. Modular or mainframe oscilloscopes accept plug-in options that provide capabilities other models may have as built-in options. Plug-in options include curve tracers, dual tracing, differential amplifiers, and high-gain amplifiers. Many probes are available for oscilloscopes, with the most common probes being ×1 and ×10 high-impedance or divider probes.

Voltage is measured with a scope by connecting the signal to be measured to the vertical input, adjusting the vertical range to produce the largest on-screen waveform, counting the number of divisions of waveform height or beam displacement, and multiplying that number by the vertical range. Frequency is measured by adjusting the horizontal range to produce the widest single cycle on the screen, counting the number of divisions for one cycle, and multiplying that number by the horizontal range.

The phase relationship between two signals of the same frequency can be determined by connecting one signal to the vertical input and the other signal to the horizontal input. With the horizontal range control on volts, a 1:1 elliptical pattern is obtained. The sine of the phase angle is determined by dividing the center height of the ellipse by its total height. Dual-trace operation offers an alternative method of phase angle measurement. Furthermore, this method gives the user the advantage of determining whether the base angle is positive or negative.

BIBLIOGRAPHY

Color comes to digital oscilloscopes, Electronic Design, December 8, 1983.

Digital oscilloscopes will dominate by 1990, T. Costlow, Electronic Design, January 12, 1984.

Storage oscilloscopes, Chris Everett, EDN, April 5, 1984.
Use of the Dual-Trace Scope, a Programmed Text, Charles H. Roth,
  Jr., Prentice-Hall, 1982.
The XYZ's of Using a Scope, Tektronix, Inc., 1983.

QUESTIONS

  1. Name and describe seven functional sections of an oscillo-scope.

  2. Describe the performance of the heater, the cathode, the control grid, the anodes, and the deflection plates in the operation of a CRT.

  3. Identify and compare the differences between standard and storage oscilloscopes.

  4. Describe the purpose and operation of vertical and horizontal amplifiers of scopes.

  5. Describe the purpose and operation of sweep circuits of scopes.

  6. Name and describe two power supply voltage needs for oscilloscopes.

  7. Name and describe four common oscilloscope specifications.

  8. Name and describe four common oscilloscope options.

  9. Name and describe the operation of the basic vertical and horizontal controls of a scope.

  10. Explain the purpose, the options, and the operation of trigger controls.

  11. Describe the voltage measurement process, including calibration, zeroing, range selection, and magnitude determination.

  12. Describe the frequency measurement process, including calibration, range selection, period determination, and frequency calculation.

  13. Under what circumstances might you use an oscilloscope with the vertical and/or horizontal channel(s) uncalibrated?

  14. Explain why the vertical channel may need recalibration.

  15. What are the primary advantages an oscilloscope has over a voltmeter?

  16. Describe two methods for measuring the phase relationship between voltages.

  17. Name two considerations you should be aware of when you are using a ×10 probe.

  18. Give two reasons for using a ×10 probe.

19. Compare chopped and alternate sweeps.

20. What considerations must you keep in mind when you are viewing a high-frequency signal by using a chopped sweep?

21. Describe a situation when you might need to use an external trigger signal.

PROBLEMS

1. Calculate the deflection sensitivity of a CRT for $V_a = 500$ volts, $\ell = 2.5$ centimeters, $L = 20$ centimeters, and $d = 1$ centimeter.

2. Calculate the deflection sensitivity of a CRT for $V_a = 750$ volts, $\ell = 2$ centimeters, $L = 18$ centimeters, and $d = 1.5$ centimeters.

3. What peak-to-peak and rms voltage is indicated by the oscilloscope trace in Figure 8–28 if the scope is on 50 millivolts per division? If it is on 200 microvolts per division?

4. What is the DC component of the waveform described in Problem 3 if the signal has AC and DC combined and the 0-volt reference is the bottom line?

5. What peak-to-peak and rms voltage is indicated by the oscilloscope trace in Figure 8–29 if the scope is on 2 volts per division? If it is on 500 microvolts per division?

6. What is the DC component of the waveform described in Problem 5 if the signal has AC and DC combined and the 0-volt reference is the bottom line?

7. What period and frequency are indicated by the oscilloscope trace in Figure 8–28 if the scope is on 20 milliseconds per division? If it is on 250 microseconds per division?

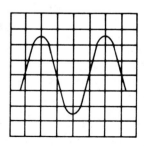 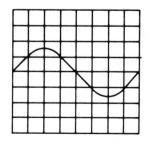 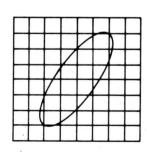

FIGURE 8–28   **FIGURE 8–29**   **FIGURE 8–30**

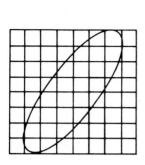 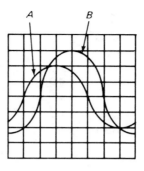 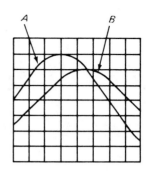

FIGURE 8–31   **FIGURE 8–32**   **FIGURE 8–33**

8. What period and frequency are indicated by the oscilloscope trace in Figure 8–29 if the scope is on 50 microseconds per division? If it is on 4 milliseconds per division?

9. What phase angle is indicated by the oscilloscope trace in Figure 8–30?

10. What phase angle is indicated by the oscilloscope trace in Figure 8–31?

11. What phase angle is indicated by the oscilloscope trace in Figure 8–32?

12. What phase angle is indicated by the oscilloscope trace in Figure 8–33?

9

Oscillators and Generators

OBJECTIVES

This chapter discusses some of the instruments that produce pulsing and alternating voltage test signals. Types of equipment, their outputs, and their specifications are described, along with typical applications.

Describe the output characteristics and applications of an oscillator.

Describe the output characteristics and applications of an RF generator.

Describe the output characteristics and applications of a function generator.

OSCILLATORS

By one definition, an *oscillator* is a *circuit* with positive feedback that produces a sine wave. By another definition, an *oscillator* is an *instrument* that produces sine waves over a wide frequency range and is used for circuit or system evaluation. Both definitions are correct, and both will be discussed in this chapter. The discussion begins with the basic oscillator circuit.

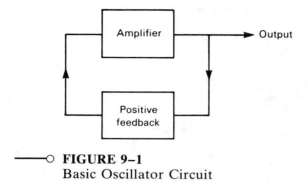

───○ **FIGURE 9–1**
Basic Oscillator Circuit

Basic Oscillator Circuits

Some people say that if you want an oscillator, you should build an amplifier. While this statement is not totally true, you will find, in your experience as a technician, that amplifiers oscillate independently against your intentions. The output of a typical amplifier is usually larger than the input and is often 180 degrees out of phase with it. If a portion of the output is fed back to the input, as indicated in Figure 9–1, and shifted another 180 degrees in the process, it can reinforce the original input and cause the amplifier to oscillate. When circuits are designed as oscillators, this positive feedback or regeneration is produced in a variety of ways.

Figure 9–2 shows a traditional oscillator circuit, the *phase shift oscillator*. Signals entering the inverting (−) input are amplified and shifted 180 degrees before they leave. The noninverting (+) input is grounded. When the output is fed back to the *RC* network, it is shifted in phase an amount that depends on the *RC* time constant.

One particular frequency for this circuit causes the phase shift of the total network to be 180 degrees. Since the input is now in phase with the output, the circuit oscillates. The frequency for oscillation is determined in the following example.

○ **Example of Frequency of Oscillation in Phase Shift Oscillator**

Using the *R* and *C* values given in Figure 9–2, we can calculate the frequency of oscillation as follows:

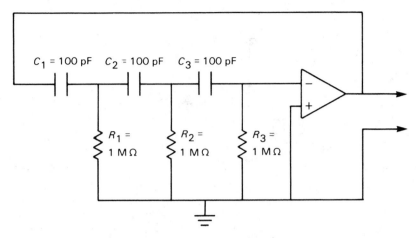

——○ **FIGURE 9–2**
Phase Shift Oscillator

$$f = \frac{1}{2\pi RC\sqrt{6}} = \frac{1}{2\pi \times (1 \times 10^{6}\ \Omega) \times (100 \times 10^{-12}\ \text{F}) \times \sqrt{6}}$$
$$= 649.7\ \text{Hz}$$

Changing each capacitor to 200 picofarads would change the frequency as follows:

$$f = \frac{1}{2\pi RC\ \sqrt{6}} = \frac{1}{2\pi \times (1 \times 10^{6}\ \Omega) \times (200 \times 10^{-12}\ \text{F}) \times \sqrt{6}}$$
$$= 324.9\ \text{Hz}$$

The *Wien bridge oscillator* shown in Figure 9–3 is widely used to produce high-quality sine voltages of frequencies up to 1 megahertz. It utilizes a differential amplifier and a feedback bridge network. One side of the bridge is a resistive voltage divider returning a portion of the output voltage to the inverting input of the amplifier. The other side of the bridge consists of series and parallel RC circuits that return a portion of the output to the noninverting input of the amplifier. Note that $R_1 = R_2$ and $C_1 = C_2$.

This circuit initiates and sustains oscillation whenever the positive feedback through the RC branch (the noninverting input) exceeds the negative feedback through the resistive branch (the inverting input). Oscillation occurs at the frequency where X_C equals R. This frequency is changed in variable-frequency oscillators by changing the RC values with a range switch and using variable capacitors.

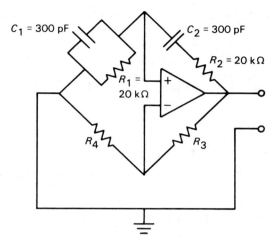

○ **FIGURE 9–3**
Wien Bridge Oscillator

○ **Example of Frequency of Oscillation in Wien Bridge Oscillator**

The frequency of oscillation for the circuit in Figure 9–3 can be calculated as follows:

$$f = \frac{1}{2\pi R_1 C_1} = \frac{1}{2\pi \times (20 \times 10^3 \, \Omega) \times (300 \times 10^{-12} \, F)} = 26.53 \, kHz$$

If both capacitors were increased to 500 picofarads, the frequency would change as follows:

$$f = \frac{1}{2\pi R_1 C_1} = \frac{1}{2\pi \times (20 \times 10^3 \, \Omega) \times (500 \times 10^{-12} \, F)}$$

$$= \frac{1}{2\pi \times 1 \times 10^{-5}} \, Hz = 15.9 \, kHz$$

Frequency Standards

A *frequency standard* is an instrument that produces a high-quality, stable signal of known frequency. As a technician, you will probably use standards built around quartz crystal oscillators. The high Q of a quartz crystal, greater than 10^6, makes it resonate over a very narrow frequency range. Since quartz crystals are temperature-sensitive, they are usually mounted in thermally insulated ovens within the frequency standard.

For even higher quality, a cesium atomic-beam frequency standard is available. It provides an accuracy of one part in 10^{13}. Rubidium gas oscillators have nearly the same degree of quality, and, like the cesium beam device, they are used as primary standards.

Variable-Frequency Oscillators

Variable-frequency oscillators, which produce sinusoidal voltages over a wide frequency range, are often designed around the Wien bridge circuit. Frequency is controlled by range-switched capacitors and variable rheostats.

Another circuit used as a variable-frequency oscillator is the *phase-locked loop* (PLL). As shown in Figure 9–4, it consists of a voltage-controlled oscillator (VCO), a phase detector, a low-pass filter, and a DC amplifier. The VCO is usually designed around a varactor or unijunction transistor, and it produces a sinusoidal signal whose frequency is controlled by a DC input voltage. This output signal enters a phase detector and is compared with the phase of an externally generated, crystal-controlled reference frequency. The output of the phase detector is a DC voltage with magnitude and polarity proportional to the size and direction of error. This signal then receives a power boost from the amplifier and is sent on to the VCO, where its output is corrected. When a PLL circuit is in the locked-in condition, its output is as stable as its crystal reference signal.

Sweep oscillators provide two simultaneous outputs. One output is a test voltage (signal A), which varies in frequency over a

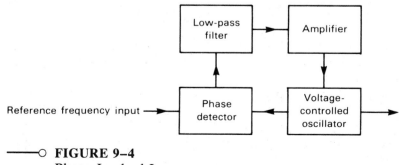

───○ **FIGURE 9–4**
Phase-Locked Loop

preselected range while maintaining a constant amplitude. The other output (signal B) is a voltage that varies in amplitude, proportional to the frequency of signal A.

In a typical application, the varying-frequency signal A is applied to the input of the test circuit, and the circuit output is connected to the vertical input of an oscilloscope. Signal B is connected to the horizontal input of the oscilloscope. Vertical deflection represents circuit output, and horizontal deflection represents frequency. The result on the CRT is a frequency response curve showing output voltage as a function of frequency.

Specifications

Common variable-frequency oscillator specifications are voltage and frequency range, accuracy, hum and noise, and percentage of distortion. Like the specifications of other instruments, these specifications vary according to price, but they tend to improve each year. The available frequency range extends from well below 1 hertz to above 10 gigahertz—not, however, in one instrument. Frequency accuracy is typically 1% to 3% of indicated value, but lower values are available.

Output voltages are adjustable with potentiometers in some instruments and with calibrated attenuators in others. Voltage range extends from millivolts to 10 volts or so. Distortion is less than 1.0%, with hum and noise less than 0.1%. Quite often, these oscillators have an output impedance of 600 ohms to match the impedance of common transmission lines. The output is usually balanced, but it can be grounded, according to the needs of the user.

Quartz frequency standards generally provide outputs in decade steps, such as 0.1, 1, 5, and 10 megahertz. Accuracy is typically five parts in 10^9 after a 15-minute warm-up. Because crystals drift slowly over a long period of time, long-term drift or frequency stability is generally specified. Frequency drift is likely to be less than five parts in 10^{10} per day.

Applications

Generally, a variable-frequency oscillator is used to evaluate the frequency response of a circuit or a system. In this test, a sinusoidal signal is applied at the circuit input, and circuit output voltage is measured. This procedure is repeated for many frequencies. Then, output voltage or gain (output voltage divided by input voltage) as a function of frequency is plotted on semilog paper. An illustration of such a graph is given in Figure 4–6. The following examples show how we determine voltage and power gains in such tests.

○ Example of Calculating Voltage Gain

Consider an amplifier with an output of 6 volts pp and an input of 20 millivolts pp at 1 kilohertz. The voltage gain A_V of this amplifier, expressed in decibels (dB), is calculated in the following manner:

$$A_V = 20 \log_{10} \frac{V_{out}}{V_{in}} = 20 \log_{10} \frac{6 \text{ V pp}}{0.02 \text{ V pp}} = 20 \log_{10} 300 = 49.5 \text{ dB}$$

If your calculator does not have a logarithm function, the value can be determined as follows: Express the gain in scientific notation—that is, as a number between 1 and 10 times a power of ten. In this case, you write $300 = 3.00 \times 10^2$. The first part of the logarithm is the exponent, 2. The second part of the logarithm follows the decimal point and is found in a table like the logarithm table in the Appendix. Find the sequence of numbers 3.00 in the logarithm table. The table indicates that the second part of the logarithm is .4771. Therefore, the logarithm of 300 is 2.4771. Here are some other examples of numbers and their logarithms:

Number	Logarithm
25	1.3979
4000	3.6021
125	2.0969

In some cases, power gain A_P is desired rather than voltage gain. The formula is as follows:

$$A_P = 10 \log_{10} \frac{P_{out}}{P_{in}}$$

○ Examples of Calculating Power Gain

The power gain for an amplifier with an input of 10 milliwatts and an output of 2.5 watts is calculated as follows:

$$A_P = 10 \log_{10} \frac{P_{out}}{P_{in}} = 10 \log_{10} \frac{2.5 \text{ W}}{0.01 \text{ W}} = 10 \log_{10} 2500$$

$$= 10 \times 3.3979 \text{ dB} = 33.98 \text{ dB}$$

The power gain for an amplifier with an input of 20 milliwatts and an output of 400 milliwatts is as follows:

$$A_P = 10 \log_{10} \frac{P_{out}}{P_{in}} = 10 \log_{10} \frac{400 \, mW}{20 \, mW} = 10 \log_{10} 20$$

$$= 10 \times 1.3010 \, dBm = 13.01 \, dBm$$

In frequency response tests, highly accurate or precise frequencies are not always critical. What is more important is the stability of the input voltage. Quite often, the varying impedance of a circuit under test produces a variable load as the oscillator frequency changes. If you are not certain of an instrument's stability, you should continually monitor its output voltage as you change frequency.

Figure 9–5 shows the test circuit for amplifier evaluation. The input meter monitors the level to ensure that it remains constant. The output oscilloscope monitors both the level and the possible distortion. The scope trace ensures that the characteristics you report are characteristics of the circuit under test, not of the source. Remember, also, to establish one common ground.

GENERATORS

Adding one more capability to the variable-frequency oscillator gives additional versatility. This added capability is modulation. *Modulation* is the combination of a radio frequency (RF) signal, called a

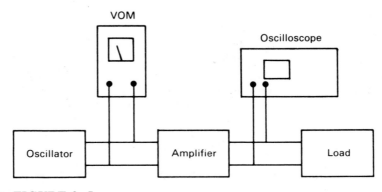

—O **FIGURE 9–5**
Amplifier Testing

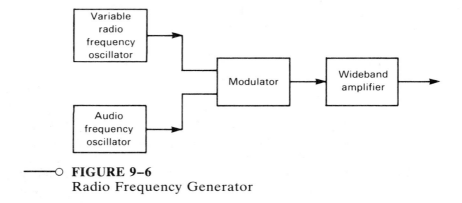

○ **FIGURE 9–6**
Radio Frequency Generator

carrier, and an audio frequency (AF) signal. An oscillator with modulation capabilities is called a *generator*. The RF generators are used to test and align a wide range of communication and navigation systems.

Radio Frequency Generators

Figure 9–6 outlines the essential sections of a radio frequency generator. One section contains a variable RF oscillator with independent frequency and amplitude controls. Another section contains an AF oscillator with an independent amplitude control. The audio frequency is usually 400 hertz or 1 kilohertz and is not variable, although the option of an external source is common.

These two signals, the RF and AF signals, are combined in a mixer (or modulator), and a composite signal is generated. The manner of mixing determines the type of output. The two most common methods are amplitude modulation (AM) and frequency modulation (FM). Figure 9–7A shows the typical internal AF signal, and Figure 9–7B shows the internal RF signal. The two common output signals are an AM signal, shown in Figure 9–7C, and an FM signal, shown in Figure 9–7D. Phase modulation (PM) is another common option.

An AM signal may also be described in terms of modulation percentage. The next example shows the calculation.

○ Example of Calculating Modulation Percentage

Consider the waveform in Figure 9–8. For this waveform, the modulation percentage is as follows, where V_{max} represents the maximum voltage and V_{min} represents the minimum voltage:

$$\% \text{ modulation} = \frac{V_{\text{max}} - V_{\text{min}}}{V_{\text{max}} + V_{\text{min}}} \times 100\%$$

$$= \frac{400 \text{ mV} - 100 \text{ mV}}{400 \text{ mV} + 100 \text{ mV}} \times 100\% = 60\%$$

Changing the minimum signal to 50 millivolts pp changes the modulation percentage as follows:

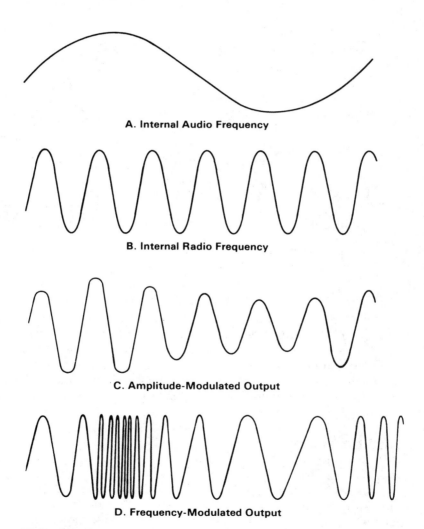

A. Internal Audio Frequency

B. Internal Radio Frequency

C. Amplitude-Modulated Output

D. Frequency-Modulated Output

○ **FIGURE 9–7**
RF Generator Signals

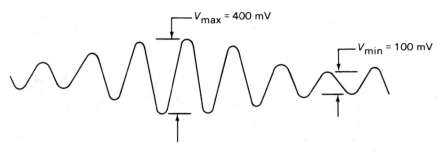

———○ **FIGURE 9–8**
Waveform for Calculating Modulation Percentage

$$\% \text{ modulation} = \frac{V_{max} - V_{min}}{V_{max} + V_{min}} \times 100\%$$

$$= \frac{400\,\text{mV} - 50\,\text{mV}}{400\,\text{mV} + 50\,\text{mV}} \times 100\% = 77.8\%$$

An FM signal may also be described in terms of a modulation index M, which is a function of the frequency deviation f_d of the carrier and the frequency f_m of the highest modulating signal. The modulation index is expressed as follows:

$$M = \frac{f_d}{f_m}$$

The generator shown in Figure 9–9 is an RF signal generator with an output frequency range from 450 kilohertz to 520 megahertz. Amplitude and frequency modulation can be added to the basic signal at controlled and monitored levels. These instruments often have other capabilities such as AM/FM, AM/AM, and FM/FM stereo. That is, separate signals are provided for dual-channel systems.

Specifications

Most RF generators offer a wide band of frequencies, with some reaching as high as 10 gigahertz. Frequency accuracy varies with cost and can range from 3% to better than 0.001% of the indicated frequency. Output amplitude is generally controlled by variable and stepped attenuators and indicated with analog or digital displays.

────○ **FIGURE 9–9**
FM/AM Signal Generator (Photo courtesty of Boonton Electronics Corporation)

Other important specifications are the control and indication of modulation percentage. The accuracy of both modulation frequency and amplitude is likely to be about 1%.

Since modulation develops frequency harmonics, harmonic content is often specified. One way of specifying harmonic content is with the term total harmonic distortion (THD). Total harmonic distortion is generally less than 1% with a laboratory signal generator. Many signal generators have PLL circuits that allow the variable-frequency output source to be locked to an internal crystal oscillator for increased accuracy and stability at selected frequencies.

Applications

Radio frequency generators find application in a variety of situations. They are used in the alignment of radios and televisions and in the calibration of radio communication and navigation devices and systems. Radio frequency generators also can be used to test antennas and transmission lines for gain, signal-to-noise ratio, and bandwidth. In summary, RF generators are used for test, measurement, alignment, and calibration.

FUNCTION GENERATORS

The continuing increase in circuit types and complexities requires a wide range of test signals. At one time, only sinusoidal signals were used for circuit evaluation. Now, nonsinusoidal and nonsymmetrical signals are needed, and their source is a *sweep function generator* such as the one shown in Figure 9–10. Each of these signals can be mathematically described as a function of time—hence the name, function generator. The range of waveforms available will be described shortly; a discussion of their mathematical representations is inappropriate for this text and is left to another course.

Circuits

The oscillators and signal generators described earlier derive their signals from *RC* circuits, such as a Wien bridge oscillator. In other words, their signal begins as a sine wave. Function generators often use other approaches because of the variety and complexity of their output signals. Even the most basic function generator offers sine, square, and triangle waveforms. Generally, then, the signal does not derive from a sinusoidal oscillator.

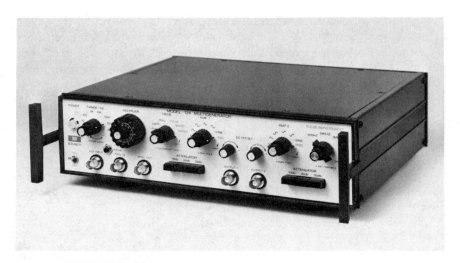

───○ **FIGURE 9–10**
Sweep Function Generator (Photo courtesty of Exact Electronics, a division of Dynatech Nevada, Inc.)

For instance, the system illustrated in Figure 9–11 begins with a triangle waveshape. This triangle waveshape is then amplified, attenuated, and sent on as output, if that is the waveshape desired. If a square or sine wave is desired, the triangle waveshape is sent to an appropriate shaping circuit; then, it is amplified, attenuated, and sent on as output. A DC offset voltage can be added if needed (more about this topic will be given later).

The triangle generator circuit, which is shown in Figure 9–12, produces a linear ramp signal. Operation centers around a capacitor that is charged from one of two constant-current sources. An electronic switch connects a positive source for a period of time, switches to the negative source, switches back to the positive source, and so on. A triangle waveform such as the one shown in Figure 9–13 is produced. Since the source provides constant current rather than constant voltage, the waveform is linear.

This signal is then amplified by a DC amplifier and sent on to the output or to the shaping circuits. It may also be sent on to a level detector. By using a comparator, the level detector senses when the upper and lower trigger levels are reached. A *Schmitt trigger,* which is a circuit that produces square waves from varying signals, then

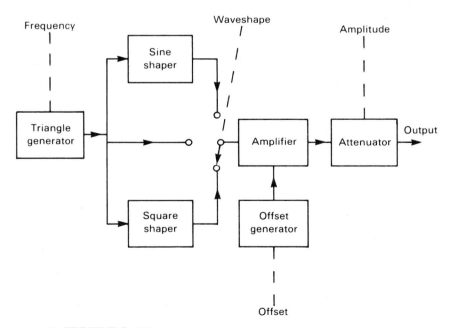

○ **FIGURE 9–11**
Triangle Function Generator

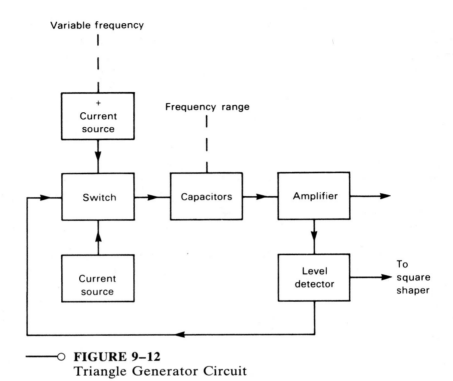

—O **FIGURE 9–12**
Triangle Generator Circuit

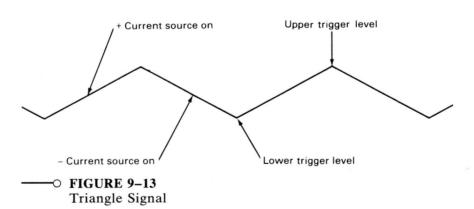

—O **FIGURE 9–13**
Triangle Signal

provides the square wave that commands the electronic switch. This signal also becomes the basis for the eventual square-wave output.

The frequency band of this generator is controlled by ranging, switching capacitors. The specific frequency within a band is controlled by potentiometers, which determine the portion of the constant current that goes to the capacitor and the portion that bypasses through a shunt.

As just noted, the square-wave output signal originates at the Schmitt trigger. Its frequency is controlled by the triangle generator. Since the square shaper begins with a square wave, little more may need to be done now. A quick-response flip-flop and wideband amplifier may be used here, though. That is, early square-wave generators began with sine waves, clipped them with diodes, and amplified the remaining signal. Even though the sine waves may have been clipped to 10% of their original height and amplified back up, they had long rise times. The Schmitt trigger and flip-flop combination produces fast switching between specific positive and negative limits and, hence, fast-rising square waves.

When a sine wave is desired, it is derived from the original triangle. Figure 9–14 shows how the sine shaper produces a sine wave from the original triangle waveform. A sine shaper circuit, such as the one shown in Figure 9–15, works as follows: The circuit ap-

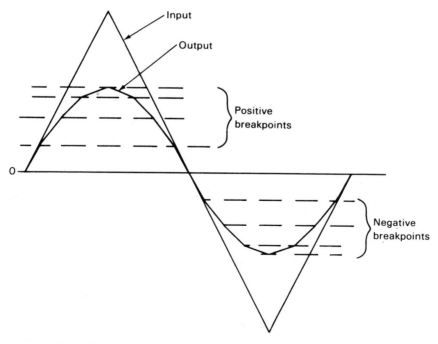

FIGURE 9–14
Sine Shaper Waveforms

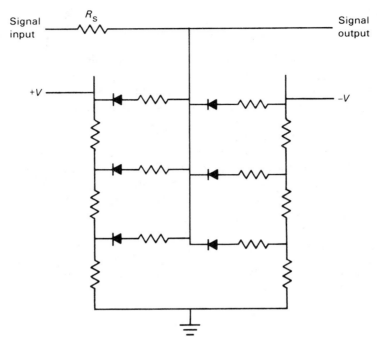

─────○ **FIGURE 9–15**
      Sine Shaper Circuit

pears as a nonlinear load to the triangle generator. The six diodes
are biased at different voltage levels. Three diodes are biased in one
polarity, and three are biased in the other polarity. At low voltage
levels, the triangle wave is unaffected. As the voltage level of the
triangle wave rises, it forward-biases one of the diodes, which, in
turn, shunts the original circuit and loads it, decreasing the slope of
the original rising waveform. This action occurs at the first positive
breakpoint, as shown in Figure 9–14. As the input voltage continues
to rise, the second diode turns on, and then the third. Each of these
actions further reduces the slope of the original waveform, changing
it to a near-sine shape.

The diodes turn off in reverse order as the input voltage de-
clines. Other diodes, biased in the opposite direction, produce a set
of negative breakpoints. The combination of both circuits converts
a triangle waveform to a sine waveform. While the resulting output
waveform is not an ideal sine wave, it approaches one. It can be
improved by adding more diodes to provide more breakpoints.

The sine shaper circuit is followed by a wideband DC amplifier
and an attenuator. A DC level control adds bias to the waveform to

offset it above or below 0 volts DC, as desired. That is, 0 volts need not be in the center of a sine, square, or other waveform. An *offset voltage* added to a sine wave can move the waveform up or down, with 0 volts being at the center, at the top of the peak, above the peak, near the bottom peak, and so on. It can produce a waveform much like that of an amplifier, a small sine wave on top of a DC bias.

The preceding discussion presented the minimum capabilities of a function generator. Most function generators, such as the one in Figure 9–10, provide many more output options. Figure 9–16 presents those options most frequently available.

Figure 9–16A shows a typical zero-centered sine signal with equal positive and negative peaks. With the addition of a positive- or a negative-bias voltage, this signal can move above or below zero to the point where the upper or the lower peak is at zero or even beyond, as shown in Figure 9–16B. The square and triangle signals in Figures 9–16C through 9–16F also have these options. Function generators offer precise frequency, amplitude (peak value), and bias (offset) control with the sine, square, and triangle signals.

Pulse signals, such as the one shown in Figure 9–16G, can be positive or negative and are adjustable in amplitude. In addition, the pulse width (duration of a given pulse) and repetition rate (number of pulses per second) can be precisely selected. Sawtooth voltages, as in Figure 9–16H, are adjustable in frequency, polarity, and amplitude. A tone burst, such as the one shown in Figure 9–16I, can have a preselected frequency, amplitude, and duration. The pulse burst, shown in Figure 9–16J, is adjustable in pulse width, amplitude, pulse rate, and duration.

The output signals described here represent a composite of offerings from a variety of function generators. Thus, all generators do not offer all these signals, and all signals offered have not been described. For example, some generators can sweep a selected frequency range by using sine, square, or triangle signals. However, the preceding description gives the essence of function generator capabilities.

Specifications

Function generator specifications can be placed into three categories: signal options, types of control, and quality of output. Typical output signal options were outlined above. The control and quality of output will be described here.

The frequency of the sine, square, and triangle signals can usually be selected with decade and vernier controls. The range is from 0.001 hertz to over 20 megahertz. Typical accuracy is under 2%.

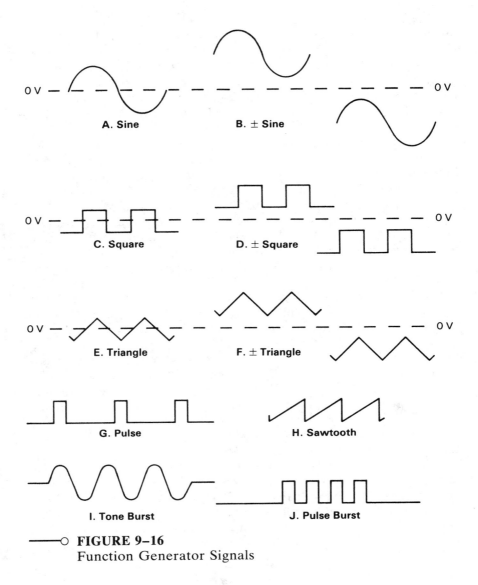

Function Generator Signals

Amplitude is generally selected with decade and vernier controls, with levels up to 20 volts pp and with an accuracy also under 2%. Long- and short-time stability for frequency and amplitude may also be specified.

Other common specifications are the linearity of triangle waveforms and the rise time, fall time, overshoot, and ringing of the pulse output signals. Output impedance is usually 50 ohms.

Applications

Function generators find use in a wide range of applications. For example, technicians and engineers develop circuit parameters by trial and error with the aid of the variety of test signals available from a generator. By trying different values, they can optimize the performance of a circuit before the design becomes final.

Function generators may also be used to evaluate the performance of complete systems before they are connected to their intended input source. In this case, anticipated input signals are simulated so that the system response can be evaluated. Finally, function generators can simulate input signals while you troubleshoot a circuit.

ADDITIONAL GENERATORS

The oscillators and generators just described represent a major portion of the AC signal sources. There are, however, many other instruments of that type. The instruments described in this section will give you an indication of other signals and operating features that are available.

Pulse Generators

While a function generator can often provide pulses, a pulse generator is designed specifically for this purpose. As Figure 9–17 shows, its circuit begins with an *astable multivibrator,* a device with two transistors that turn on and off, producing a square wave. This part of the circuit provides a number of pulses with a rate controlled by the repetition rate control. These pulses then trigger a *monostable multivibrator,* a transistor-switching circuit that produces one pulse for each input command. The duration of this pulse is controlled by the pulse width control.

A typical resulting output waveform is shown in Figure 9–18. The monostable multivibrator in this case has an on-time (T_1) of 0.25 millisecond. This time is the pulse width. With the astable multivibrator operating at a rate of 800 hertz, it produces a pulse with a period T_2 calculated as follows:

$$T_2 = \frac{1}{f_2} = \frac{1}{800 \text{ Hz}} = 1.25 \text{ ms}$$

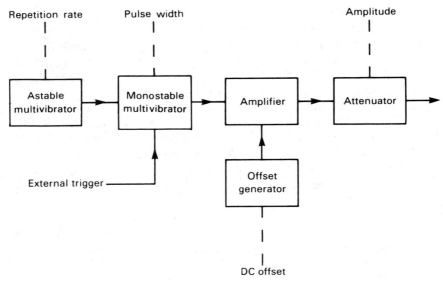

───○ **FIGURE 9–17**
Pulse Generator

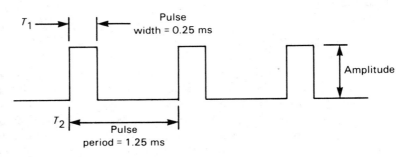

───○ **FIGURE 9–18**
Pulse Waveforms

Another parameter of a pulse waveform is its *duty cycle*. Its calculation is illustrated in the following example.

○ **Example of Calculating Duty Cycle**

The duty cycle of the waveform in Figure 9–18 can be calculated in this way:

$$\text{duty cycle} = \frac{\text{pulse width}}{\text{pulse period}} \times 100\%$$

$$= \frac{T_1}{T_2} \times 100\% = \frac{0.25 \text{ ms}}{1.25 \text{ ms}} \times 100\% = 20\%$$

Changing the pulse width to 0.10 millisecond also changes the duty cycle, as the next calculation shows:

$$\text{duty cycle} = \frac{T_1}{I_2} \times 100\% = \frac{0.10 \text{ ms}}{1.25 \text{ ms}} \times 100\% = 8\%$$

Programmable Function Generators

The combination of synthesizers and microprocessors in programmable function generators expands the already extensive offerings of function generators. As mentioned earlier, a synthesizer can develop a wide range of high-quality test signals. With a microprocessor, elaborate test routines and sequences can be programmed for automatic execution. Then, the routine can be stored in memory for recall when the need for it recurs. A programmable function generator is shown in Figure 9–19.

Most programmable instruments use keyboards rather than rotary or push-button switches, thereby reducing switch or control noise. This arrangement also allows users to interface with remote control or to monitor systems for automated test data transmission

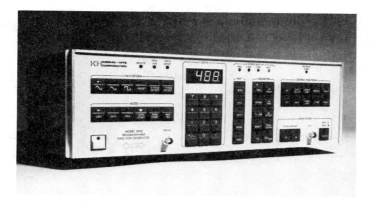

○ **FIGURE 9–19**
Programmable Function Generator (Photo courtesy of Krohn-Hite Corporation)

and analysis. As more instruments become programmable with microprocessors, technicians will need to understand programming and interfacing techniques.

Random-Noise Generators

One major concern in the field of communications is signal-to-noise (S/N) ratio. Ideally, a signal that is received or processed should be much higher than the accompanying noise. The design of RF and microwave communication systems involves the development of amplifiers and filters that discriminate between noise and signal. Random-noise generators are used in this process; they provide random electric noise of known magnitude.

Random-noise generators or sources are used in conjunction with noise figure meters. Noise figure, also called signal-to-noise ratio, describes the ratio between a typically small signal as received and the electric components added to it by circuitry. Causes of noise figure include variations in active-component temperature and inadequate shielding.

In a typical test, the random noise generator provides a known magnitude of noise at the circuit input, and the meter measures the noise at the output. The technician's task is to shield, filter, amplify, or perform whatever design modifications are necessary in order to remove the noise while allowing the signal to pass.

Television Test Generators

Both broadcast and cable television have created a demand for specially designed test and measurement instruments. A typical test signal generator can provide the synchronization, color, convergence, and chroma signals needed for the calibration and alignment of cameras and other video equipment. Other generators produce composite television signals that are sent down a transmission cable for the alignment of distribution amplifiers. Generally, these instruments are dedicated; that is, they are installed in an equipment rack at the studio or other point of distribution.

SUMMARY

Oscillators and generators are as common as multimeters, oscilloscopes, and power supplies in electronics shops and laboratories. Their outputs can range from a single-frequency sine wave to a wide band of complex waveforms. Equipment selection depends on the signal shape and quality requirements of the application. Oscillators and generators can be used for calibration, alignment, troubleshooting, and circuit design analysis.

The heart of many oscillators is a Wien bridge oscillator. This circuit consists of a differential amplifier connected to an RC bridge, which phase-shifts positive feedback and directly connects negative feedback. Oscillation occurs at the frequency where the bridge resistance and reactance are equal. Frequency can be changed by changing the R and the C values.

A frequency standard is an accurate, precise, and stable frequency source. An accuracy of one part per 10^{13} is possible with rubidium gas oscillators, making these instruments among the highest-quality standards for any parameter. Quartz standards are used in most laboratories and can provide an accuracy of five parts per 10^9.

Test oscillators are used for testing the frequency response of circuits and offer a wide frequency range, generally from 1 hertz to 1 megahertz. Amplitude and frequency accuracy in these instruments is likely to be between 1% and 3%. Sweep oscillators offer the same type of output but with the added feature of automatic frequency sweeping. A second output signal provides a linear voltage proportional to the frequency of the first output. This added signal is used to horizontally sweep an oscilloscope display while the response of a test circuit is connected to the vertical input. This device produces an automatic frequency response curve.

A generator is an oscillator with modulation capabilities. Radio frequency generators often have ranges from below 1 hertz to over 500 megahertz, with frequencies to 10 gigahertz available. Outputs can be phase-, amplitude-, or frequency-modulated, with combinations of AM and FM also possible. Alignment and calibration of communication and navigation systems are common applications.

Function generators produce a variety of waveforms, including sine, square, triangle, pulse, and bursts. These waveforms are available over a frequency range from 0.001 hertz to 20 megahertz, with amplitude and frequency accuracies of 2% or better. Function generators can often sweep a wide frequency range with constant or varying amplitudes. Typical applications include network and system analysis, troubleshooting, and design optimization.

Programmable function generators combine signal synthesizers and microprocessors to produce high-quality complex signals with known and controlled frequencies. Microprocessor control makes the function generator compatible with automated control, monitoring, and analysis systems. Random-noise generators produce noise of known magnitude for use in the design of RF and microwave communication systems, amplifiers, and filters. Television test generators produce synchronization, color, chroma, and other signals of controlled magnitude for the test and alignment of amplifiers, transmission lines, and other components of television systems.

BIBLIOGRAPHY

Signal generator operates between 1 and 1040 megahertz, T. Costlow, Electronic Design, October 27, 1983.

Advances forecast in crystal oscillators, P.J. Klass, Aviation Week & Space Technology, November 28, 1983.

What is a function generator? Robert Baetke, Test & Measurement World, November, 1983.

Sine generator has low distortion. A.D. Helfric, EDN, January 26, 1984.

Fundamentals of Electronics, Douglas R. Malcolm, Jr., Breton Publishers, 1983.

Operational Amplifiers for Technicians, Jefferson C. Boyce, Breton Publishers, 1983.

QUESTIONS

1. Name the basic components of a Wien bridge oscillator, and describe what they do.

2. What controls the frequency of a Wien bridge oscillator? Of a crystal oscillator? Of an atomic frequency standard?

3. Explain the difference between an oscillator and a generator.

4. Describe the output and the application of a random-noise generator.

5. Compare the characteristics and specifications of an oscillator with those of a frequency standard.

6. Describe the output characteristics and specifications of a sweep oscillator.

7. Name and describe a typical application for a sweep oscillator.

8. Describe the difference between the way an oscillator produces a sinusoidal waveform and the way a function generator produces one.

9. Describe how a function generator produces a square wave.

10. Explain the differences between sine, + sine, and − sine waveforms.

11. Name and describe four common output signals provided by function generators.

12. Describe the use for the second output of a sweep oscillator.

13. Explain how a VCO circuit can be controlled from a remote source.

14. When you are using a variable-frequency oscillator to test a circuit, why should you consider 1 kilohertz to be midway between 100 hertz and 10 kilohertz?

15. What is the primary difference between a square-wave generator and a pulse generator?

16. What advantage does a multivibrator have over a sinusoidal oscillator–diode clipper circuit for producing square waves?

17. Name and describe four output characteristics of a pulse waveform.

18. Describe a pulse burst.

19. Describe a tone burst.

PROBLEMS

1. Determine the frequency of the phase shift oscillator in Figure 9–2 if R = 10 kilohms and C = 100 picofarads.

2. Determine the frequency of the phase shift oscillator in Figure 9–2 if R = 15 kilohms and C = 500 picofarads.

3. Determine the frequency of the phase shift oscillator in Figure 9–2 if R = 100 kilohms and C = 400 picofarads.

4. Determine the frequency of the phase shift oscillator in Figure 9–2 if R = 470 kilohms and C = 200 picofarads.

5. Determine the frequency of the Wien bridge oscillator in Figure 9–3 if R_1 = R_2 = 20 kilohms and C_1 = C_2 = 400 picofarads.

6. Determine the frequency of the Wien bridge oscillator in Figure 9–3 if R_1 = R_2 = 30 kilohms and C_1 = C_2 = 600 picofarads.

7. Determine the frequency of the Wien bridge oscillator in Figure 9–3 if R_1 = R_2 = 15 kilohms and C_1 = C_2 = 200 picofarads.

8. Determine the frequency of the Wien bridge oscillator in Figure 9–3 if R_1 = R_2 = 47 kilohms and C_1 = C_2 = 150 picofarads.

9. What is the voltage gain of an amplifier with an output of 6.6 volts pp and an input of 55 millivolts pp?

10. What is the voltage gain of an amplifier with an output of 840 millivolts pp and an input of 5.25 millivolts pp?

11. What is the power gain of an amplifier with an output of 5 watts and an input of 100 milliwatts?

12. What is the power gain of an amplifier with an output of 5 watts and an input of 50 milliwatts?

13. Calculate the modulation percentage of an AM signal whose V_{max} is 300 millivolts pp and whose V_{min} is 100 millivolts pp.

14. Calculate the modulation percentage of an AM signal whose V_{max} is 450 millivolts pp and whose V_{min} is 350 millivolts pp.

15. Calculate the modulation percentage of an AM signal whose V_{max} is 400 millivolts pp and whose V_{min} is 50 millivolts pp.

16. Calculate the modulation percentage of an AM signal whose V_{max} is 300 millivolts pp and whose V_{min} is 50 millivolts pp.

17. Determine the pulse period and duty cycle of a pulse generator with a 10-kilohertz astable multivibrator and a monostable multivibrator with an on-time of 30 microseconds.

18. Calculate the effects of changing the frequency of the astable multivibrator in Problem 17 to 20 kilohertz.

19. Determine the pulse period and duty cycle of a pulse generator with a 150-kilohertz astable multivibrator and a monostable multivibrator with an on-time of 1 microsecond.

20. Calculate the effects of changing the frequency of the astable multivibrator in Problem 19 to 200 kilohertz.

10

Electronic Counters

OBJECTIVES

This chapter presents the basic operation, specifications, and applications of digital, electronic frequency and universal counters. Techniques for frequency and time measurements and for the counting of events are described.

Name and describe parameters measured by a typical universal counter.
Describe the basic operation of a typical universal counter.
Discuss the procedures for the measurement of frequency, period, and time interval.
Name and describe four typical counter specifications.

THEORY OF OPERATION

Modes of Operation

An electronic *counter* is an instrument that counts. It counts the number of cycles in a second and displays the result as frequency. It can measure the time for one cycle and display the period, or it can measure the time interval between any two commands. It also has the capability of determining the ratio between two frequencies. Finally, events can be counted and totaled.

Figure 10–1 shows the block diagram for a basic counter. The signal whose frequency is to be measured is applied to the signal

213

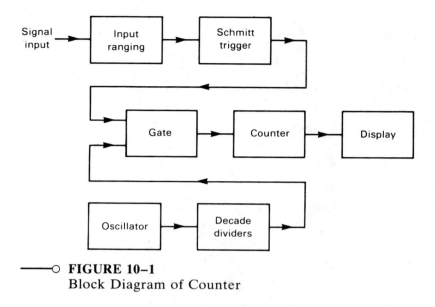

FIGURE 10–1
Block Diagram of Counter

input terminals. Then, it is amplified or attenuated, reshaped with a Schmitt trigger, and sent to the gate. Also arriving at the gate is a signal from a crystal oscillator. This signal has been reduced in frequency, by decades, to a frequency appropriate for the measurement range. It will open and close the gate, which, in turn, will cause the counter to count the pulses coming from the Schmitt trigger. Finally, the resulting count goes to the display. This process will be described in more detail shortly.

The preceding discussion gives an overview of frequency measurement or cycle counting. Counters can also count events, determine the ratio of two signals, and measure the time for one cycle. While the functions mentioned here are not the only ones available, they are the most common. And since frequency measurement is the most typical application, the description in the next subsection begins with that system.

Time Base

A counter simply counts cycles or pulses for a given amount of time. Then, it displays this count as a frequency. The task of having a counter count pulses is not difficult. What is difficult is having the counter stop counting when a specific amount of time has elapsed. In an electronic counter, this task is accomplished by a high-quality *time base,* a circuit that is the foundation for all measurements. Depending on the type of measurement being made, it provides an accurate period of time, such as 1 second, or a specific number of pulses per second, for reference.

In most laboratories, frequency and time measurements are the most accurate and precise measurements you can make. They are so precise and accurate because electronic counters determine time very well. The basis for this ability—the heart of all electronic counters—is an oscillator, typically, a 10-megahertz crystal oscillator.

The high Q of the crystal combined with a regulated voltage source produces an accurate and stable frequency, or time source. Placing the crystal in a temperature-controlled oven reduces the drift caused by the quartz temperature coefficient. The combination of these features leads to a time base with an accuracy of 1 part per million (ppm) and a long-term drift that is often less than 3 parts per million per month.

Figure 10–2 shows the block diagram for a counter's time base. The sine output of the oscillator is reshaped into a square wave by a Schmitt trigger (this operation will be described shortly) and sent on to a series of decade dividers. Each divider then produces an output square wave that is one-tenth of the frequency of the input square wave. The dividers perform their operations with cascaded flip-flops. Once the square wave is divided by 10, it is passed on to the next divider, and then it is passed on to another, and so on. Thus, if a 10-megahertz square wave enters the divider chain, it is reduced to 100 kilohertz, 10 kilohertz, 1 kilohertz, and so on, down to 1 hertz.

The number of dividers used—and, therefore, the resultant square-wave frequency—is controlled by the range switch. This output square wave is then used to operate the gate, which simply turns the counting circuit on and off. The counting duration is the period of the square wave that was selected by the range switch.

As previously mentioned, the Schmitt trigger changes a sine (or other varying) signal into a square wave. This circuit, as shown in Figure 10–3, consists of a monostable multivibrator. With a low input signal, transistor Q_1 is off, making its collector high. This action turns Q_2 on, making its collector low. A high input on Q_1 makes its collector low, turns Q_2 off, and raises the collector voltage of Q_2 up to V_{CC}. The turning on and off of Q_1 provides larger (or more profound) on-off commands for Q_2.

The output signal of the Schmitt trigger is a clearly defined square wave, as illustrated in Figure 10–4. The levels of the input signal where transition occurs are controlled by transistor biasing. These levels are adjustable in the Schmitt trigger circuits described

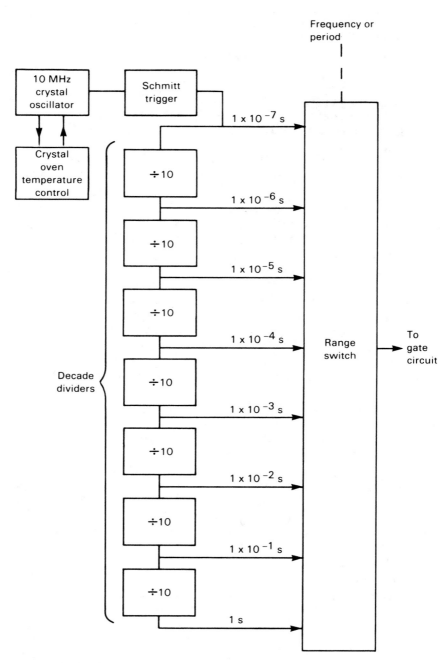

FIGURE 10–2
Time Base of Counter

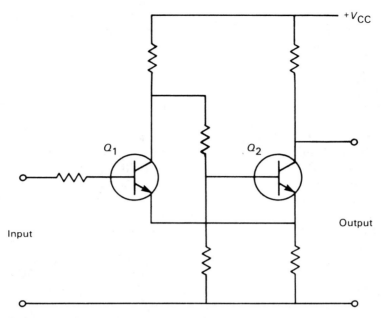

—○ **FIGURE 10–3**
Circuit for Schmitt Trigger

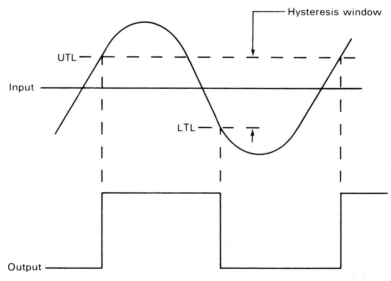

—○ **FIGURE 10–4**
Schmitt Trigger Waveforms

later in this chapter. However, this application uses upper (UTL) and lower (LTL) transistor levels equidistant from zero but in opposite directions.

Output signal transition occurs whenever the input crosses the UTL or the LTL. A signal must go up through the UTL and down through the LTL in order for a pulse to be produced. Variations that occur within the *hysteresis window* are not detected. *Hysteresis* is the word typically used to describe a dead zone such as the lag that occurs in a transformer when magnetic field reversal does not immediately follow current reversal. Mechanical gear trains have a similar problem, often called backlash. In the Schmitt trigger circuit, the hysteresis window describes the area betwen the limits where slope changes are not detected. Hysteresis is desirable here, though, because it allows the operator to selectively exclude small noise transients that would otherwise be identified as primary signal transitions.

Since the incoming signal in this case comes from an internal oscillator, it is sinusoidal and of constant magnitude. A good-quality square wave is easily maintained, and it will be used to turn the counting process on and off.

The output of the dividers is a pulse of stable duration, from 1 millisecond up to 1 second. It could be more or less with different models. It is, however, always in steps of ten (decades). This signal is used to turn the counting process on and off.

While the specific circuit for turning the counting process on and off varies from one unit to another, the approach is similar to the gate circuit shown in Figure 10–5. The square wave of the on-off command enters the clock (C) input of a flip-flop, causing it to toggle once per cycle. The flip-flop output enables or disables an AND gate. Also entering this gate are the pulses to be counted. When the flip-flop is high, the gate pulses with each input pulse. These pulses continue on for counting. When the flip-flop is low, the gate output remains at zero.

Signal Preparation

When you are measuring frequency, the signal you are measuring can be of varied size and shape. It might have high harmonic content, and what you consider one cycle may be difficult for the circuit to detect. Therefore, the signal usually needs processing.

Figure 10–6 shows the usual sequence of signal preparation in the signal input circuit. An input amplifier usually is the first device in the circuit. This amplifier produces an input impedance of 1 meg-

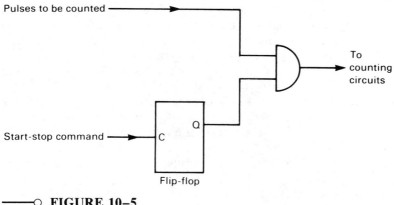

——○ **FIGURE 10–5**
Gate Circuit

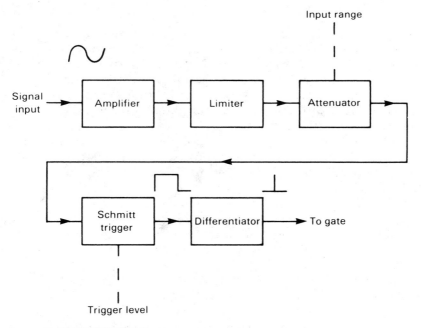

——○ **FIGURE 10–6**
Block Diagram of Signal Input Circuit

ohm or more in order to prevent loading of the source. Like the amplifier of an electronic VOM, it amplifies the input signal to a level consistent with the needs of the circuits that follow and isolates the source from the circuits that follow.

A diode limiter is the next device in the circuit, and it prevents signals that are too large from proceeding and causing damage. Remember that *input overloading is a concern with counters*. An attenuator follows the limiter and works in conjunction with the input amplifier to produce a signal of appropriate size. The input range switch is connected with these circuits.

Once the input signal is of workable size, it is forwarded to a Schmitt trigger, which gives it a workable shape. Like the Schmitt trigger in the circuit of the time base, it changes a varying signal into clearly defined pulses. The difference here is that the incoming signal can have any shape. So, a critical adjustment to be made at this point by the operator is trigger level. This adjustment is described in the following subsection.

Triggering

The waveform in Figure 10–7 has some harmonic content. When the UTL point is set low, as for Figure 10–7A, the fundamental frequency dominates and is measured. However, setting the UTL too high, as for Figure 10–7B, causes a harmonic to switch the trigger. Since this signal has multiple slope changes within each fundamental cycle, it could be difficult to detect the limits of one cycle. Thus, trigger level is a critical adjustment.

Additional signal definition is accomplished by sending the modified signal through a differentiator (see Figure 10–6). The differentiator produces a sharp positive pulse on the rise of the square wave and a sharp negative pulse on its fall. Counting the positive pulses reduces even further the possibility of a miscount.

In summary, a simple sine or a complex input waveshape is reshaped by the counter's circuitry into discrete pulses, one for each cycle of the input signal. They are now ready to be counted.

Counting and Display

Pulses are usually counted by four-bit, *decimal-counting units* (DCUs), utilizing the 8421 binary-coded decimal (BCD) format. Figure 10–8 shows an example of a DCU. Pulses to be counted enter the clock (C) input of the first flip-flop. The output of this flip-flop represents the least significant bit (LSB), or 1s, described in binary as 2^0. This output terminal is called Q, and the signal proceeds to the clock input of the next flip-flop. It is also available for other circuitry.

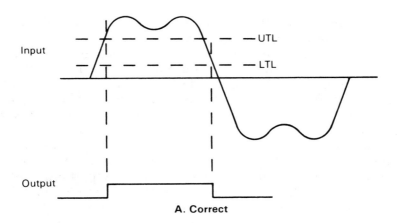

Input

UTL

LTL

Output

A. Correct

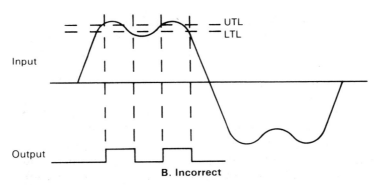

Input

UTL
LTL

Output

B. Incorrect

─────○ **FIGURE 10–7**
Trigger Levels

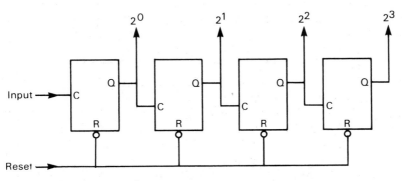

2^0 2^1 2^2 2^3

Input ──▶   C   Q   C   Q   C   Q   C   Q

R   R   R   R

Reset ──▶

─────○ **FIGURE 10–8**
Decimal-Counting Unit

The second flip-flop produces output pulses that represent 2s, or 2^1. The third flip-flop counts 4s, or 2^2, and the fourth flip-flop counts 8s, or 2^3, which is also called the most significant bit (MSB). Together, the four flip-flops produce four-bit binary numbers from 0000_2 to 1001_2, representing the decimal values from 0_{10} to 9_{10}.

Figure 10–9 shows how the four flip-flops of the DCU count 11 input pulses. The first flip-flop switches high on the trailing edge of the first input pulse. The second flip-flop switches high on the trailing edge of its input—that is, when flip-flop 1 goes low. This event occurs on the fall of the second input pulse. Flip-flop 1 alternates back and forth, going high on odd pulses, while flip-flop 2 goes high on even pulses. Flip-flop 3 goes high on the trailing edge of its input, that is, at the fall of the fourth input pulse. Flip-flop 4 goes high on the trailing edge of the eighth input pulse. Note that reset occurs on the tenth pulse.

As with all decimal counting; once the base limit (9 or 1001) is reached, a carry of 1 (or 0001) is added to the next column and the original column is reset to 0. Thus one DCU is needed for each decimal place. The decimal places may range from four to ten or more, depending on the range and the precision of the specific counter.

Figure 10–10 shows a counter with six decimal places. An incoming signal is sent to the appropriate decimal-counting unit, according to its frequency and the measurement range selected. Suppose that it goes directly to the first DCU on the right, which like all others, can count in decimal from 0 to 9. When the count reaches

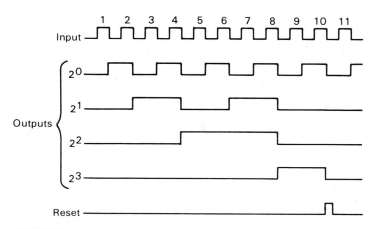

—○ **FIGURE 10–9**
DCU Waveforms

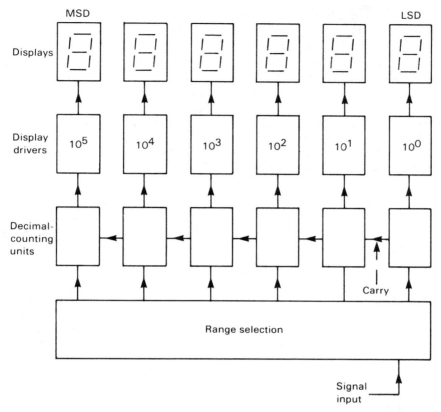

─────○ **FIGURE 10–10**
Counter Display System

9, the first DCU sends a 1 to the next DCU and recycles itself back to 0. The combination of two DCUs has counted up to 10. When the second and first DCUs reach 99, a 1 is sent to the third DCU, and the first two DCUs recycle. Thus, six DCUs can count up to 999,999. The DCU on the right is called the LSD (least significant digit), because it represents 1s. The DCU on the far left represents 100,000s, and it is called the MSD (most significant digit).

The output of each DCU is connected to a decoder–display driver, which operates the light-emitting diode (LED) or liquid crystal display (LCD) indicators; see Figure 10–10. The display drivers simply connect the necessary voltage to the combination of segments that represent the decimal number currently stored in the associated DCU. In addition to numbers, the counter may display decimal points, overranging, and plus or minus signs. Once again, the display depends on the range of capabilities of the counter.

In some counters, the increasing count is continually displayed. In other counters, only the final count is shown. Regardless of what is displayed, though, the count continues until the gate closes, which, in many cases, is at the end of 1 second. At this point, the total count of input cycles is displayed. Since the count was conducted for 1 second, the count represents the frequency, or number of cycles, in 1 second. Whether or not the counter recycles and counts again, and how soon it does, is controlled by the display time and reset controls, which will be described shortly.

SPECIFICATIONS AND OPTIONS

Electronic counters are categorized as frequency counters or universal counters, depending on their range of capabilities. Frequency counters generally measure only frequency, while universal counters measure frequency and various other parameters. Many counters can also measure pulse width, measure rise and fall times, and perform mathematical data analysis. A typical electronic counter is pictured in Figure 10–11.

Specifications

The specifications of a universal counter describe its capabilities and how well it can perform. Specifications also provide a way to compare counters and select the best one for a given situation. As always, the user should prioritize the specifications to determine which ones are most important. The following paragraphs describe common specifications and their typical ranges.

The frequency range of most counters is from less than 1 hertz to 100 megahertz. This range is extended above 50 gigahertz in some counters, and it can be extended in others with the addition of plug-in adapters.

While accuracy is generally expressed as a function of time base error, trigger error, and resolution, an overall accuracy of 1 part per million in a six-digit result is a reasonable value to expect. Values to 1 part in 10^8 are not uncommon. A more accurate result can be obtained by following the manufacturer's instructions.

Time base accuracy is the foundation of a counter's quality and is expressed in short-term and long-term stability. Typical short-term

—————○ **FIGURE 10–11**
Electronic Counter (Photo courtesy of Hewlett-Packard
Company)

values are 1 part per 10^6 to 1 part per 10^{10} per second. One part per
10^8 per month is a typical long-term stability specification.

Functions for electronic counters include measurement of the
frequency and the period of signals. In addition, universal counters
can perform two variations. First, they can totalize—that is, count
the number of events in a predetermined amount of time. Second,
they can measure the time interval between stop and start commands. Some counters can measure the ratio between two input
frequencies. Finally, some counters can scale, or divide, an input
frequency and produce a square-wave output that is a fraction of the
input frequency.

Options

Some counter options are built-in, and others are available as plug-
in units. The most common option is a frequency converter, or pre-
scaler, which extends the counter range upward. With this option, a
100-megahertz counter could become capable of measuring frequencies into the gigahertz range. High-pass and low-pass input filters
remove undesired signals or noise before it gets to the input.

Additional displays can indicate amplitude, heterodyne offset frequency (the signal mixed internally by a counter to scale an incoming frequency down to measurable size), or the results of data analysis. With the use of an internal microprocessor, data analysis, such as mean, minimum, maximum, and standard deviation, can be automatically performed. Data can be made available as a digital output. Internal computer interface circuitry is also becoming more common.

APPLICATIONS

The most common uses for a universal electronic counter are the measurement of frequency and time and the counting of events. Most counters are simple to operate and interpret since they have few controls and utilize digital displays. The primary concern is the magnitude of the input signal. If the signal is too small, it can be difficult to detect. On the other hand, a signal that is too large can damage the front end of the instrument. As always, you should know the capabilities and limitations of any instrument you plan to use.

Controls and Adjustments

Most counters have a function switch that selects the operating modes. Generally, these modes include frequency, period, count, manual start, and manual stop. In the frequency mode, the counter measures the frequency of the incoming signal. It measures the time elapsed for one cycle when it is in the period mode. The count mode causes it to count incoming pulses until told to stop by external trigger signals. The instrument will begin to count time when the manual start command is given, and it will stop when the manual stop is selected. A test mode is sometimes also available. In this mode, an internal test signal should produce a predetermined count to be displayed as an instrument self-test.

The length of counting time for frequency measurement is controlled by a time, time base, or frequency switch. This switch also serves as a range selector during period measurements. An input or a sensitivity switch indicates if the input is a direct signal or a signal from a plug-in unit. It also indicates the approximate signal size that is passing through the input amplifiers and attenuators.

Display time and reset controls are used in the recounting process. When a count is completed, the result is displayed. If the display time control is on hold, the count will remain indefinitely. However, a reset command will cause an immediate recount, and that count will then be held. With a readjustment of the display time control, an operator can select the amount of time a count will be displayed before an automatic recount begins. Generally, a counter will continue to recount but display the original count until it changes.

The switches just described are the essential controls of universal counters. More sophisticated counters may have multiple inputs and outputs, synthesized programmers, microprocessor-based data analyzers, and additional measurement capabilities. However, all universal counters have some basic controls that must be adjusted. A reasonable set of beginning adjustments is as follows: Select the correct function; select a 1-second time base; select the highest input voltage range; and select a midrange display time. Of course, you should review the manual first.

Frequency Measurement

Frequency is measured by selecting the frequency mode and connecting the signal to the input terminal of the counter. A circuit similar to the circuit represented in Figure 10–12 is used. In the counter, the incoming signal is adjusted to a workable magnitude, squared, changed to pulses, and sent on to the gate. At the same

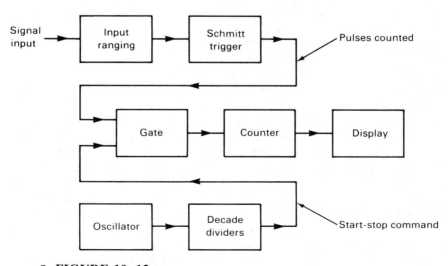

———○ **FIGURE 10–12**
Circuitry for Frequency Measurement

time, squared oscillator pulses, which have been reduced to the desired counting period, also arrive at the gate. The counter simply counts the input signal cycles for the predetermined period and displays that count on the digital readout.

Input signal magnitude is the primary concern in this measurement. It must not exceed the input limit of the counter. Once you are certain of an acceptable amplitude, select a reasonable gate time with the time base switch. As stated earlier, a time of 1 second is a good beginning value. With the display time control on midrange, a count should soon appear. If it does not, the sensitivity, or input level, control should be switched downrange.

When a count does appear, you can read it directly from the display, since most counters also display the appropriate prefixes. Your only other responsibility is to select a gate time that produces the most precise indication. At one end of the spectrum, a very high frequency can exceed the number of digits capable of being displayed. In this case, a shorter gate time is required. Usually, though, 1 second is not too long a time. For example, a 1-second gate time will produce a three-digit display of frequencies between 100 and 999.

Since most counters include an error of ±1 in the last digit, a display of 345 could mean 344, 345, or 346. Increasing the gate time to 10 seconds will produce an indication such as 345.6. This value represents a true value of 345.5, 345.6, or 345.7. So, the degree of precision is increased by one significant figure.

When the frequency you are measuring is low, a long measurement time is needed for adequate precision. In other words, measuring a 100-hertz signal for 1 second produces three significant figures, while measuring 1 megahertz produces six significant figures. One remedy is to measure 100 hertz for 10 seconds in order to obtain four significant figures or for 100 seconds to obtain five significant figures. A more reasonable solution is to measure the period of the signal. For example, a counter capable of measuring within ±1 microsecond can measure the period of a 10-hertz signal (period of 0.01 second) to five significant figures. Then, you simply calculate frequency. This method is far superior to measuring frequency to two significant figures, with one of those figures rounded off.

Period Measurement

The decision to measure period rather than frequency is not made according to which answer you need. You can always calculate one from the other. The choice depends on which method will produce the most accurate and most precise result, as indicated in the preceding discussion. Generally speaking, period is measured for low-frequency signals, and frequency is measured for high-frequency signals. Period rather than frequency is selected for period measurements, and the range switch is turned to seconds rather than hertz. Otherwise, though, the two measurement procedures are similar.

While the operator's procedures are almost the same for frequency and period measurements, the circuitry is different, as shown by a comparison of Figures 10–12 and 10–13. As indicated in Figure 10–13, in the period mode, the incoming signal controls the gate's opening and closing, while the oscillator output becomes the signal to be counted.

As an example of the difference in precision for frequency and period measurements, let us consider a 1-hertz signal. If its frequency is measured for 1 second, a one-place result is displayed. However, if the output of a 1-megahertz oscillator is counted for the period of a 1-hertz signal, a six-place result is displayed.

To an operator, then, two differences exist between frequency and period measurements. First, the period mode must be selected. Second, the result is in seconds rather than hertz.

Time Interval Measurement

The time elapsing between two events can be measured by selecting the time mode. This mode causes the counter to count the pulses of its own oscillator for the duration of time between those two events.

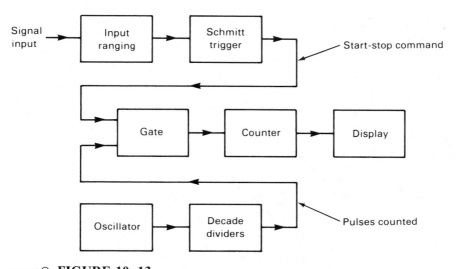

○ **FIGURE 10–13**
Circuitry for Period Measurement

As Figure 10–14 shows, start and stop commands come from external triggering sources to the controlling reset-set (RS) flip-flop. Trigger levels must be adjusted when external sources are used to ensure that triggering occurs at the desired point on the command signals. Front-panel controls operate the internal triggers.

Since the oscillator produces 1 million to 10 million pulses per second, which must be counted for time interval measurements, durations longer than 1 second could exceed the capacity of the display. For durations longer than 1 second, then, you should change the range so that decade dividers are introduced betwen the oscillator and the counter.

Ratio Measurement

The ratio mode causes a counter to determine the ratio between the frequencies of two external signals. While you could measure both and calculate the ratio, the ratio mode provides an accurate and continuous method for monitoring it.

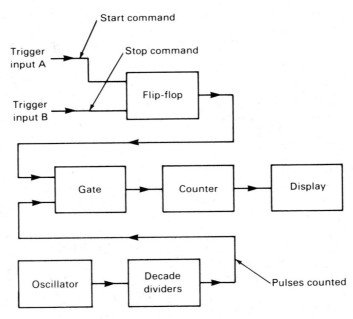

—○ **FIGURE 10–14**
Circuitry for Time Interval Measurement

A circuit like the one represented in Figure 10–15 is used for ratio measurement. This system uses two similar input preparation circuits—that is, amplifiers, limiters, attenuators, and Schmitt triggers. The higher-frequency signal is connected to the input location designated for it, and the lower-frequency signal is applied to the input designated for it. Both signals arrive at the gate as square waves. The low-frequency signal commands the gate to open, the high-frequency signal is counted, and the low-frequency signal closes the gate. The number of cycles of the high-frequency signal for one cycle of the low-frequency signal is the ratio of the one to the other. Decade dividers provide for frequency ranging.

Totalizing

The totalizing process is somewhat the same as frequency measurement in that events are counted for a period of time. However, there are differences. First, the input pulses or variation are not necessarily periodic, as they are when frequency is measured. Second, the input level may vary. Third, the input is likely to be random rather than periodic. Fourth, the desired duration of the measurement may

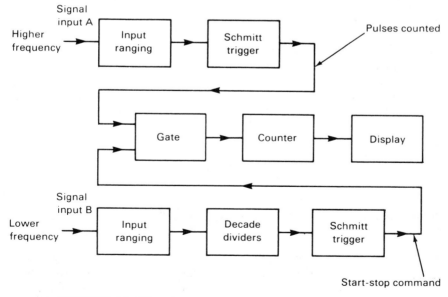

○ **FIGURE 10–15**
Circuitry for Ratio Measurement

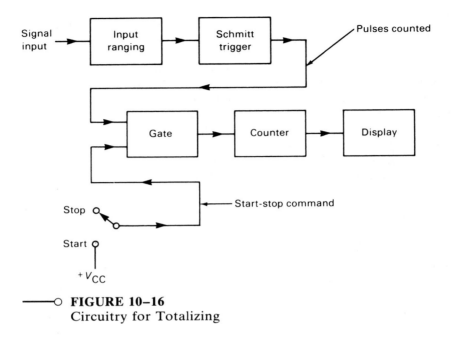

────○ FIGURE 10–16
Circuitry for Totalizing

not coincide with the internal time base but may be controlled by external or manual commands.

Because of the differences just listed, the trigger level must be adjusted so that the counter knows what to count and what to ignore. Furthermore, start and stop command voltages must be produced and connected to the gate input, as shown in Figure 10–16. Then, the instrument can be instructed to start. A current count of events will be continuously displayed, and counting will continue until stop is commanded. The stop command may be given with a front-panel switch or external sources.

SUMMARY

Electronic counters are probably the most accurate and precise instruments you will find in a typical electronics lab. This quality is derived from the counter's time base, which is designed around a quartz crystal oscillator. Typically, the crystal is located in a voltage- and temperature-controlled oven and operates at a frequency of 1 or 10 megahertz, ±1 part per million. The oscillator output sine wave is then reshaped into a square wave by a Schmitt trigger.

Decade dividers built around cascaded flip-flops reduce the oscillator output in steps of ten, to 100 kilohertz, 10 kilohertz, and so on, down to 1 hertz. One of these frequencies is then selected by the range switch and is used to turn a gate flip-flop on and off. The gate, in turn, controls the length of time that the counter counts pulses, and it does so with 1-part-per-million accuracy.

The signal to be measured is also processed on its way to the counter. First, an amplifier-limiter-attenuator circuit provides buffering to prevent source loading and changes the signal magnitude to a workable size. Second, another Schmitt trigger changes the signal to discrete pulses, a more workable shape.

The pulses continue on to be counted by the decimal-counting units (DCUs). These units use the 8421 binary-coded decimal (BCD) format and produce binary numbers from 0000 to 1001 to represent the decimal values 0 to 9. Each DCU then operates an LCD or LED display through a decoder and a driver. The number of DCUs and displays determines the resolution of the counter, which can range from as low as 4 up to 11 or more.

Specifications of a counter include frequency range, accuracy, time base accuracy, and functions. Most electronic counters can measure frequencies from DC up to 100 megahertz, with some reaching above 50 gigahertz. Accuracy is based on many factors, including time base error, trigger error, and resolution. An accuracy of 1 part per 10^8 is not uncommon. Time base accuracy is described in short-term and long-term values, with 1 part per 10^6 to 10^{10} for 1 second (short-term) and 1 part per 10^8 for 1 month (long-term) being typical values.

Frequency counters measure frequency, while universal counters measure frequency, period, and time interval and totalize events. Some counters can also determine the ratio between two signals and scale an incoming signal down in decades. More sophisticated counters have synthesized programmers for the generation of complex measurement routines. Others have microprocessor-based data analysis circuits that can determine mean, maximum, minimum, and standard deviation.

Some low-cost options, both built-in and plug-in, are also available for most counters. Such options include high- and low-pass filters, heterodyne frequency converters for extending the range upward, and multi-input adapters. Digital data output circuits and computer interfaces are two other common options.

Counter operation is not complex. You must first ensure that the signal you are measuring does not exceed the counter's frequency or amplitude limits. Then, the signal is applied to the counter input and the desired mode is selected. Frequency measurement consists of selecting the frequency mode, the highest input voltage range, and the 1-second time base. The input range may need to be reduced if the incoming signal is too small. Then, the frequency is read from the display. An important point to keep in mind is to select the time base that gives the greatest number of digits without exceeding the counter's limits.

For period measurements, the internal circuit arrangement is different from the arrangement for frequency measurement, but the external operations are similar. In this case, period is selected rather than frequency. Also, the counter displays time rather than cycles per second or hertz. The mode used, period or frequency, depends on which one will produce the most precise result, that is, the one giving the greatest number of digits in the display.

Events can be counted over a period of time, as determined by the internal time base or external start-stop trigger signals. When this mode is used, you must adjust the signal trigger level in order to ensure that the proper events are counted and smaller events are ignored. When external start-stop signals are used, other trigger controls must also be adjusted. The measurement of elapsed time involves the count of oscillator pulses for a selected duration. Once again, start-stop commands can be given by front-panel controls or external trigger signals.

BIBLIOGRAPHY

Working with counters, R. Grossblatt, Radio-Electronics, November, 1983.

Electronic Instrumentation and Measurement Techniques, 2nd ed., William David Cooper, Prentice-Hall, 1978.

Digital Electronics, Logic and Systems, 2nd ed., John D. Kershaw, Breton Publishers, 1983.

QUESTIONS

1. Name and define four parameters that can be measured by a universal counter.

2. Sketch the block diagram of a universal counter, and name all sections.

3. Describe the basic operation of each counter section, and discuss how these sections interrelate.

4. Name and describe four common specifications for universal counters.

5. Explain the process by which a counter provides its high accuracy. What degree of accuracy is to be expected?

6. Describe the role that a Schmitt trigger performs in a universal counter.

7. Describe the role of the gate circuit in a counter.

8. State the cautions that must be observed in counter use.

9. Explain the purpose of a frequency converter.

10. Describe two situations for which two input signals would be needed.

11. Explain the purpose of the display time control.

12. Describe how to measure period with a counter.

13. Describe how to measure frequency with a counter.

14. Compare the internal circuit arrangements for period and frequency measurements.

15. Explain the importance of a proper triggering level when you are measuring the frequency of complex waveforms.

16. Describe how to count events with a counter.

17. Explain how a counter could be used to measure phase.

18. Explain how a counter measures the ratio between two different frequencies.

19. Explain the role of the trigger control in totalizing events.

20. Compare the use of a counter with the use of an oscilloscope in the measurement of frequency.

11

DC Voltage
Sources

OBJECTIVES

This chapter describes applications, characteristics, and sources of DC voltage, with the primary focus on laboratory power supplies. Circuit operation, characteristics, and specifications are defined and described.

Describe the characteristics of ideal DC voltage.
Name and describe some common sources of DC voltage.
Describe the circuits and operation of the three basic sections of a DC power supply.
Describe the function of power supply regulators.
Describe the operation of two common regulator circuits.
Name and describe five DC power supply specifications.

DC VOLTAGE

Most electronic circuits operate on power from a DC voltage source, such as a battery or a power supply. Source and voltage characteristics vary greatly according to the requirements of each specific application. Our discussion begins with some uses, characteristics, and sources of DC voltage.

Uses of DC Voltage

Almost every electronic product utilizes direct current in some section of its circuitry. This DC can be the signal to be processed or the power to operate the processing circuit. In the first case, transducers produce DC voltages proportional to the magnitude of some other physical variable. A thermocouple, for example, produces a voltage proportional to temperature and allows electronic processing of temperature data. Typically, these signals are in the microvolt and millivolt range.

In contrast to analog information, which has a linear range, digital information is usually represented by two different and distinct DC voltage levels. One level represents the binary number 1 and the other represents the binary number 0. Groups of these 1s and 0s can be combined to represent letters, decimal numbers, symbols, or data transmission commands.

In the second case, DC voltage can be the source of power to process information. The voltage level used depends on the needs of the circuit. Some components operate with less than 10 volts, while others, such as CRTs, require more than 10 kilovolts. Current demands may range from milliamperes to amperes, while power requirements can be insignificant or kilowatts. In summary, DC power requirements can range from microvolts to kilovolts and from microwatts to kilowatts and can be used to represent data or to process data.

Ideal Characteristics

The ideal characteristics of DC voltage are as follows:

1. It has a magnitude that is accurately and precisely known.
2. It is absolutely smooth, with no measurable ripple.
3. It is absolutely stable, with no random variations caused by source or load effects.
4. It has 0 ohms source impedance.
5. It is portable.

The order and the degree of importance of each of these characteristics depend on the specific application of the voltage. Portability is the key factor with DC sources for VOMs, whose internal circuits can often accommodate weaknesses in other characteristics.

In voltage standards, accurate and precise knowledge of the magnitude is the most important characteristic. Here, also, other limitations such as lack of portability may not be a concern.

Ripple, which is residual AC on top of the desired DC output voltage, is undesirable. Since ripple is AC voltage, it has frequency as well as magnitude. As will be discussed shortly, *ripple frequency* is equal to the input frequency in half-wave rectifiers and is twice the input frequency for full-wave and full-wave bridge rectifiers.

The degree to which ripple is removed indicates the quality of the filter circuits. The maximum amount of ripple that remains is a common specification. Other specifications are the anticipated impacts of variations in input voltage, called *source effects,* and variations in output load, called *load effects.*

Sources of DC Voltage

The most common source of DC voltage is a battery. At one time, there were just two types to choose from, a zinc-carbon (dry) primary cell or a lead-acid (wet) secondary cell, the latter being rechargeable. Today, a wide range of batteries is available. Disposable alkaline batteries are commonly used as low-cost power sources in portable equipment. Nickel-cadmium (NiCd) cells are used when current demands may be high and a rechargeable source is desired. Unlike the earlier lead-acid cell, a nickel-cadmium cell can be tipped without concern about spilling.

Lithium-based primary cells are often found in small instruments such as watches and cameras since they offer high energy density (watthours per kilogram). Lithium–sulphur dioxide cells have a higher current capacity than lithium-iodine cells. Both have a long shelf life. The decision to use a lithium-based, a lead-acid, a nickel-cadmium, or an alkaline cell depends on application needs, including voltage, current, environment, stability, and life expectancy.

Another common source of DC voltage is the line-operated power supply. It can be either an off-the-shelf, universal model, adapted to the specific application, or a custom-made model. Once again, the needs of the application will dictate the specifications of the source.

Two additional categories of DC voltage sources should be mentioned here since they are important to the field of measurement. The first is the DC voltage standard, which provides precise and accurate magnitudes of DC voltage for instrument calibration. It was described in Chapter 2. The second category is the laboratory power supply, used primarily to operate experimental circuits. Its description, which begins with a review of basic power supply concepts in the next section, is the focus of the remainder of this chapter.

DC POWER SUPPLIES

A typical DC power supply converts line frequency alternating current to one or more desired levels of DC voltage. This section reviews the transformers, rectifiers, and filters that produce a smooth DC voltage of the desired level.

Transformers

Figure 11–1 shows the essential sections of a simple DC power supply: transformer, rectifier, and filter. Another section, the regulator, will be discussed later in this chapter.

The *transformer* provides the following three functions in a power supply:

1. It isolates the load from the power line ground.
2. It changes the voltage level from the line level to one or more magnitudes.
3. It provides center taps for circuits that require them.

The turns ratio of a transformer describes the ratio of primary to secondary voltages. The turns ratio is as follows:

$$\frac{V_{secondary}}{V_{primary}} = \frac{turns_{secondary}}{turns_{primary}} \quad \text{or} \quad \frac{V_{sec}}{V_{pri}} = \frac{n_{sec}}{n_{pri}}$$

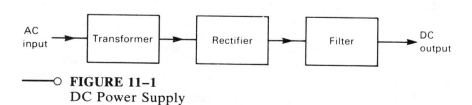

—○ **FIGURE 11–1**
DC Power Supply

○ **Example of Turns Ratio**

Consider a situation where you have a 118-volt rms source and you want 100 volts peak at the secondary. The turns ratio is as follows:

$$\frac{100 \text{ V peak}}{118 \text{ V rms} \times 1.414} = 0.6 = \frac{0.6}{1}$$

A transformer with a turns ratio of 0.6:1 would be required. That is, the secondary would have 0.6 times as many turns as the primary. This transformer would be followed by a rectifier.

Rectifiers

The basic purpose of a *rectifier* is to convert AC voltage to pulsating DC voltage. Filters that follow the rectifiers make the voltage smooth. Figure 11–2 shows three basic rectifier circuits: half-wave (Figure 11–2A), full-wave (Figure 11–2B), and full-wave bridge (Figure 11–2C). Rectifiers consist of a transformer to change the voltage to the correct magnitude, one or more diodes to perform the rectification, and the resistance of the load, whether applied internally by design and/or externally by an operator. Descriptions and comparisons of these circuits are given in the following paragraphs.

For this discussion, we will consider the secondary voltage of the power transformer to be 200 volts pp at 60 hertz, as shown in Figure 11–3A. When this voltage is applied to the half-wave rectifier of Figure 11–2A, it forward-biases the diode D_1 for one-half of the input cycle and reverse-biases it for the other half. That is, when the top of the transformer is positive and the bottom is negative, D_1 is forward-biased. Thus, D_1 conducts, and it produces an output waveform across load resistor R_L similar to the first half of the input waveform. When the input waveform goes through its second half cycle, the top of the transformer is negative and the bottom is positive. In this case, D_1 is reverse-biased, it does not conduct, and the voltage across R_1 is zero. One cycle of the output voltage waveform is shown in Figure 11–3B.

The full-wave rectifier circuit in Figure 11–2B has the same transformer secondary voltage of 100 volts peak. However, only one-half of the secondary is connected at one time. During the first 180 degrees of the voltage cycle, the transformer top is positive and the bottom is negative. Diode D_1 is forward-biased through R_L to the center tap, while D_2 is reverse-biased. The 50 volts peak of the upper

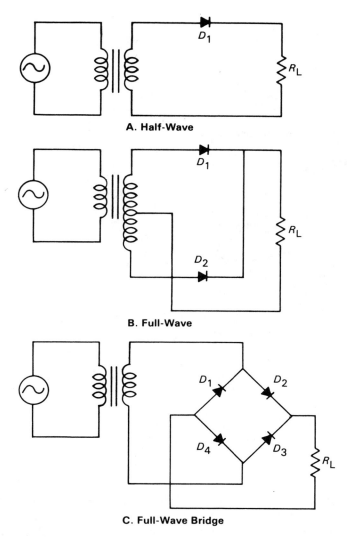

A. Half-Wave

B. Full-Wave

C. Full-Wave Bridge

──O **FIGURE 11–2**
Rectifier Circuits

half of the transformer causes D_1 to conduct through R_L for the
second half of the cycle, since the polarity of the transformer secondary reverses. One cycle of the output waveform for this circuit
is shown in Figure 11–3C.

The full-wave bridge rectifier in Figure 11–2C utilizes the complete transformer secondary voltage all of the time. Diodes D_4 and

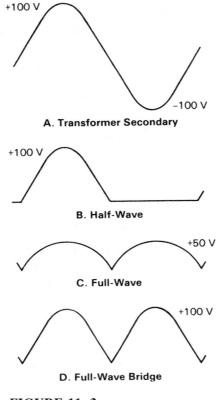

A. Transformer Secondary

B. Half-Wave

C. Full-Wave

D. Full-Wave Bridge

○ **FIGURE 11–3**
Rectifier Waveforms

D_2 conduct through R_L during the first half of the input cycle. Diodes D_1 and D_3 are reverse-biased and off. During the second half of the input cycle, D_1 and D_3 conduct through R_L, and D_2 and D_4 are off. One cycle of the output waveform for this circuit is shown in Figure 11–3D.

○ Example of Calculating Voltage for Rectifiers

Let us calculate the voltage for each of these three rectifier circuits. For the half-wave rectifier, the DC voltage is as follows:

$$V_{DC} = \frac{0.637 \times V_{peak}}{2} = \frac{0.637 \times 100\ \text{V}}{2} = 31.9\ \text{V DC}$$

Multiplying the peak by 0.637 determines the average value. This value is divided by 2 since there is conduction only during one-half of each cycle.

For the full-wave rectifier, the DC voltage is as follows:

$$V_{DC} = 0.637 \times \frac{V_{peak}}{2} = 0.637 \times \frac{100\ V}{2} = 31.9\ V\ DC$$

In this case, the peak value is divided by 2 because only one-half of the transformer is used at a time.

For the full-wave bridge rectifier, the DC voltage is as follows:

$$V_{DC} = 0.637 \times V_{peak} = 0.637 \times 100\ V = 63.7\ V\ DC$$

These three rectifier circuits can be summarized and compared as follows:

Circuit	Output (Volts DC)	Ripple (Volts pp)	Ripple Frequency (Hertz)
Half-wave	31.9	100	60
Full-wave	31.9	50	120
Bridge	63.7	50	120

The higher DC output voltage of the full-wave bridge is one obvious advantage of this circuit. Two others are a lower magnitude of ripple voltage and a higher ripple frequency, both of which make filtering easier.

Filters

The purpose of a *filter* is to remove impurities. In a power supply, the filter is used to remove—or at least reduce—the ripple. The three circuits to be discussed here are shown in Figure 11–4: capacitor (Figure 11–4A), inductor (Figure 11–4B), and pi (Figure 11–4C). We will consider the input of the filter to be from the full-wave rectifier just described.

The voltage applied to the capacitor in the capacitor filter (Figure 11–4A) has a peak value of 100 volts, and its waveform resembles the waveshape shown in Figure 11–5A, before the filter is applied. In operation, the capacitor charges up to the peak value at the 90-degree point as the rectifier diodes conduct. With no load

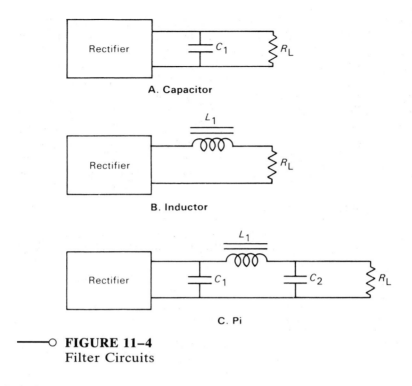

A. Capacitor

B. Inductor

C. Pi

——○ **FIGURE 11–4**
Filter Circuits

(R_L), the capacitor would remain at that voltage level, since it would have no path in which to discharge. However, the circuit does have a load R_L, and so the capacitor does discharge at a rate equal to the time constant $(R_L \times C_1)$. The output waveform for one cycle is shown in Figure 11–5B.

Figure 11–4B shows an inductor filter, with inductor L_1 in series with the load R_L. Once again, the opposition to change helps the filtering process. In contrast to the capacitor, which was in parallel and opposed change in voltage, the inductor is in series and opposes change in current. Its effect is also expressed by a time constant, L_1/R_L. Figure 11–5C shows that the inductor reduces the output ripple and causes it to lag in input by up to 90 degrees and to center around the average ($0.637 \times$ peak) magnitude of the input voltage.

The pi filter (Figure 11–4C) is a complex series-parallel circuit. Hence, a description of its operation is beyond the scope of this book. We have included the pi filter and its waveform (Figure 11–5D) here simply for completeness, since technicians often encounter this configuration in their work.

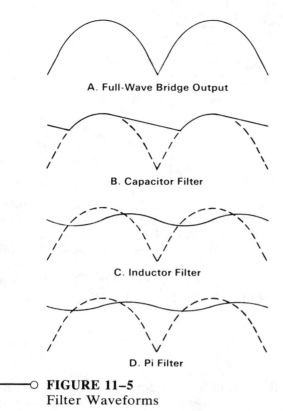

A. Full-Wave Bridge Output

B. Capacitor Filter

C. Inductor Filter

D. Pi Filter

○ **FIGURE 11–5**
Filter Waveforms

One way to examine the pi filter is to consider the oppositions. Inductor L_1 in series with the load provides a small resistance to DC with considerable reactance to AC. Capacitors C_1 and C_2 provide low-reactance shunt paths to AC while offering high parallel resistance to the load R_L.

Each of these reactive components has its advantages. The capacitor raises the DC voltage level and filters best with a light load—that is, when R_L is large and the RC time constant is long. An inductive filter improves as the load increases, because as R_L becomes smaller, the L/R time constant becomes longer. The pi filter combines the advantages of capacitors and inductors.

The quality of all of these filters is expressed by the *ripple factor* of the output, as follows:

$$\text{ripple factor} = \frac{V_{AC\ rms}}{V_{DC}} \quad \text{and} \quad \%\ \text{ripple} = 100\% \times \text{ripple factor}$$

○ **Example of Ripple Calculation**

Consider a power supply with 100 millivolts of ripple and an output of 70 volts DC. The percent ripple is as follows:

$$\% \text{ ripple} = \frac{0.100 \text{ V rms}}{70 \text{ V DC}} \times 100\% = 0.14\%$$

REGULATORS

In this section, we will examine two common power supply *regulators,* the voltage regulator and the current regulator. The schematic for a typical regulator is shown in Figure 11–6. As indicated in the schematic, a regulator monitors the output (voltage or current) and maintains it at a constant level. As will be shown in the following subsections, a regulator has many components. Thus, it is the most complex stage of the laboratory power supply.

Voltage Regulators

The most common type of power supply regulator is the voltage regulator. It maintains the output at a predetermined level regardless of variations in the power line (source) or in the demand placed on the output (load).

Regulation is defined as a percentage and can be determined as follows:

$$\text{regulation percentage} = \frac{V_{NL} - V_{FL}}{V_{NL}} \times 100\%$$

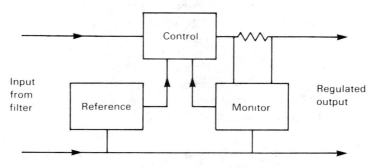

──○ **FIGURE 11–6**
Regulator

where the subscript NL represents "no load" and the subscript FL represents "full load."

○ **Example of Calculating Regulation Percentage**

Consider a power supply that is adjusted for 10 volts (V_{NL}) with no load applied but the output drops to 9 volts (V_{FL}) when the load is applied. Then, the regulation percentage is calculated as follows:

$$\text{regulation percentage} = \frac{V_{NL} - V_{FL}}{V_{NL}} \times 100\%$$

$$= \frac{10\text{ V} - 9\text{ V}}{10\text{ V}} \times 100\% = 10\%$$

The heart of most voltage regulators is a zener diode. As Figure 11–7 shows, a zener diode conducts well when it is forward-biased ($+V$), but conducts only in microamperes when it is reverse-biased ($-V$). However, when the magnitude of reverse voltage exceeds the

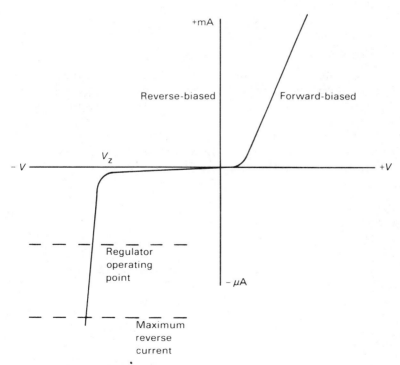

─────○ **FIGURE 11–7**
Zener Diode Characteristics

breakdown level, the diode conducts very well. This breakdown voltage (V_z) is both predictable and stable. In normal regulator operation, a resistor is connected in series with this diode to limit the current flow to one-half of the maximum reverse current ($I_{z\,max}$).

○ Example of Voltage Regulator

Figure 11–8 shows a typical application of a shunt voltage regulator. Consider the zener diode to have a breakdown voltage V_z of 15 volts and a maximum current $I_{z\,max}$ of 50 milliamperes. Since the load is unknown, R_2 will be used to provide some load on the circuit. In this design, R_2 will carry an amount of current equal to that of the zener diode. Consider also that the incoming voltage is approximately 40 volts. The calculations are as follows:

$$I_z = \frac{I_{z\,max}}{2} = \frac{50\,\text{mA}}{2} = 25\,\text{mA}$$

$$R_2 = \frac{V_2}{I_2} = \frac{V_z}{I_z} = \frac{15\,\text{V}}{25\,\text{mA}} = 600\,\Omega$$

$$V_1 = V_T - V_z = 40\,\text{V} - 15\,\text{V} = 25\,\text{V}$$

$$R_1 = \frac{V_1}{I_1} = \frac{V_1}{I_2 + I_z} = \frac{25\,\text{V}}{50\,\text{mA}} = 500\,\Omega$$

In this circuit, D_1 will maintain a 15-volt drop as long as the input remains well above 40 volts. If the input voltage rises, the diode conducts more, causing an increased voltage drop across R_1. When the input voltage drops, the diode conducts less and V_1 decreases. If a voltage less than 15 volts is desired, a divider can be used after the regulator. Typically, however, the output is not taken directly from the output of a shunt regulator like this one. Instead,

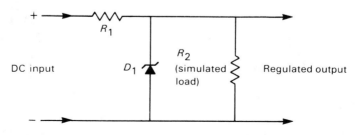

── ○ **FIGURE 11–8**
Shunt Voltage Regulator

the shunt regulator is used as the reference, and additional circuitry is used for control.

One of the most common voltage regulator circuits is the series pass regulator, which is shown in Figure 11–9. Once again, a zener diode is used as the reference voltage. However, the load current passes through Q_1, the controlling transistor. As mentioned earlier, an increase in power supply load current tends to decrease output voltage. However, this decrease increases the emitter-to-base voltage V_{BE} of the transistor, since V_z is a constant. The increase in V_{BE} causes the transistor to conduct more, which, in turn, increases the output voltage. The voltage is calculated as follows:

$$V_{out} = V_z + V_{BE} \quad \text{or} \quad V_{BE} = V_{out} - V_z$$

A transistor conducts more by a decrease in the emitter-to-collector resistance. This dynamic resistance is in series with the load. In other words, a drop in output voltage causes a reduction of the resistance in series with it and, in turn, an increase in the output voltage. An increase in output voltage causes the opposite effect. While this circuit can have a higher current limit than the shunt regulator, its range is limited by the capacity of the series transistor.

A switching regulator such as the one shown in Figure 11–10 controls the output voltage by rapidly switching the output current on and off. Transistor Q_1 is turned on by the operational amplifier when the feedback reference voltage (negative input) drops below the reference level (positive input). Since this system detects small changes, the transistor output is a square wave with a ripple frequency of up to 20 kilohertz. This ripple is easily filtered by a small inductor L_1 and a small capacitor C_1. Since the transistor is either

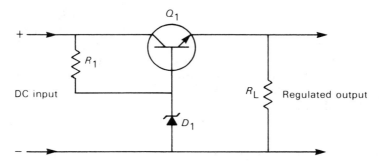

——○ **FIGURE 11–9**
Series Pass Voltage Regulator

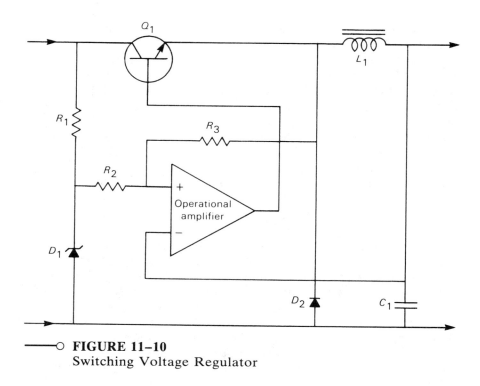

———○ **FIGURE 11–10**
Switching Voltage Regulator

fully on or fully off, it has a low power loss, making this device an efficient regulator.

Current Regulators

Current is regulated for two specific reasons: to maintain a specific amount and to limit the maximum. Figure 11–11 shows a series voltage regulator with current limiting. As before, an operational amplifier compares two signals, the desired output from reference D_1–R_3 and the actual output from divider R_1–R_2. The difference, or error, between these voltages determines the output of the operational amplifier. This amplifier then increases or decreases the current flow of transistor Q_1.

Resistor R_5 is in series with the load and produces a voltage drop in proportion to the load current. When this current reaches a specific magnitude, it causes transistor Q_2 to turn on and its collector to be near ground potential. This circuit action brings the base of Q_1 to the same point, turns Q_1 off, and interrupts the output current flow.

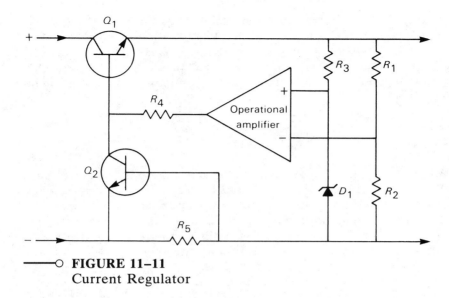

○ **FIGURE 11–11**
Current Regulator

LABORATORY POWER SUPPLIES

A laboratory power supply, such as the one shown in Figure 11–12, can be found at almost every electronics technician's workbench. It might be used as an adjustable source of DC voltage for energizing

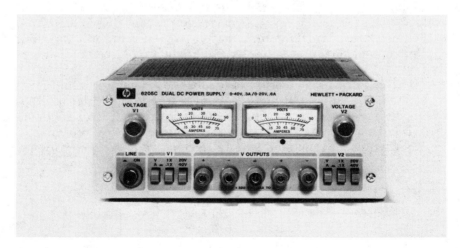

○ **FIGURE 11–12**
Dual-Output DC Power Supply (Photo courtesy of Hewlett-Packard Company)

experimental circuits or as a temporary substitute for the built-in supply of a system being tested. Applications can vary according to the voltage, capacity, and other specifications required. The specifications of these units will be discussed in the next section. This section presents basic circuit operation, common controls, and characteristics.

Basic Circuits

Figure 11–13 shows the essential stages of a laboratory power supply. At the first stage, four functions are generally performed. Line power is turned on or off, the supply is fused from the source, the supply is isolated from the line ground, and the incoming voltage is stepped up or down. A transformer performs the last two steps. Isolation reduces the danger of shock or damage from crossed grounds.

The second stage is a rectifier circuit, which changes the incoming AC line voltage to pulsating DC voltage. This rectifier is generally a full-wave bridge rectifier made of semiconductor diodes. The third stage is a filter that changes the pulses to a smooth DC voltage. Inductors and capacitors are used here; the inductors are in series with the load and the capacitors are in parallel. Specific values depend on the anticipated load current and the acceptable level of ripple.

In laboratory power supplies, the filter is followed by a regulator stage, which has the responsibility of maintaining the output voltage at a preselected level. It may also prevent the voltage from exceeding a present maximum, it may control current, or it may limit the maximum current. A few typical regulator circuits were described in a preceding section.

The essence of regulation is feedback. A desired output level is selected, the actual output level is monitored, any difference between desired output and actual output creates an error, and the error is fed back to an earlier stage to command the required change. In other words, a regulated supply automatically compensates for

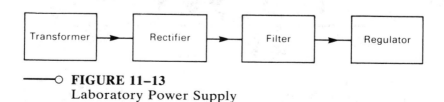

─────○ **FIGURE 11–13**
Laboratory Power Supply

changes in line voltage (source effect) or changes in the load demand (load effect). In addition to voltage regulation, power supplies often have current regulation. In this case, preselected levels of current can be maintained.

Common Controls

A power on/off switch controls the connection of the line power to the supply. The load on/off switch controls the connection of the output voltage to the terminals. With the load switch off and the power switch on, power can be removed from a test circuit with the supply still energized. This technique helps to maintain supply stability. Input and output fuses or circuit breakers are generally located on the rear of the cabinet.

Output voltage is generally selected with a voltage level control and measured with an internal voltmeter. Sometimes, calibrated decade switches instead of a potentiometer are used for voltage selection. A current level control sets the desired current flow when current regulation is desired or sets the maximum current limit when the unit is under voltage control. The output voltage appears at the two terminals marked (+) and (−); a third terminal is marked GND, for ground. The operator selects either positive or negative ground, according to the needs of the situation.

Operation

The method of operating a power supply varies greatly from one model to another. However, there are some procedures common to all units. For example, always connect the load before the supply is energized. Also, energize the load with a minimum voltage first; then, raise it to the desired level. This procedure prevents supply overload or other damage if the circuit is shorted or otherwise misconnected. Use an external voltmeter if you are not sure of the accuracy of the internal indicators of a supply. Finally, *never remove the energized power supply leads from a test circuit without first de-energizing the power source.*

Grounding is an issue of extreme importance. While many parts of an electronic circuit may have common points, there can be only one point where all common points are connected to ground. If more than one ground exists, small currents called ground loops will flow, causing interference signals in sensitive circuits. When you are using more than one power supply, such as positive and negative sources, only one supply can serve as ground. Once again, *there must be only one ground return point in a power supply system.* The preferred grounding point is the primary power source.

Characteristics

Most laboratory power supplies weigh under 10 pounds, have a volume under 1 cubic foot, and are designed for bench-top use. While many have handles and can be moved from one place to another, they are not considered to be portable. That is, most supplies cannot tolerate rough handling. Most solid-state supplies require no warm-up time, but some older models have a specified warm-up time before they reach stabilization.

   The selection of a power supply is an important decision. Selection depends on the needs of the situation, which can be determined with a few questions. What voltage is needed (what is the range)? What accuracy is needed (what are the metering and calibration)? How smooth must the voltage be (how much filtering is needed)? How stable must the voltage be (what regulation is required)? What polarity is needed (can there be a floating ground)? Proper selection requires knowledge and understanding of power supply specifications.

SPECIFICATIONS OF LABORATORY POWER SUPPLIES

Power supplies can be described by many specifications, including output voltage, physical size, cabinet style, and cost. Here, we will consider only those specifications that describe the range and quality of the output voltage. These factors are the ones that determine if the supply will cost hundreds or thousands of dollars. While dozens of parameters can be considered, the specifications that are usually of most interest are output voltage, load effect, source effect, and periodic and random deviation. Each of these parameters will be discussed in turn.

Output Voltage

The voltage specification describes the range over which the output level can be adjusted. While many possibilities exist, typical values are 0–10, 0–50, and 0–500 volts. The dual-range supply in Figure

11–12 offers a user the choice of either 0–20 or 0–40 volts, selected by a front-panel switch.

Accuracy and resolution specifications describe how accurately and precisely a desired voltage can be attained. Accuracy values can range from more than 3% to less than 0.1%. Resolution can be less than 0.05%. Current and power ratings indicate the maximum current and power capacity of a supply. Power capacity can range from 10 to 200 watts or more.

Load Effect

The load effect specification describes the degree to which changes in supply load affect the output voltage level. Also called load regulation, it is calculated in this way:

$$\% \text{ load effect} = \frac{V_{NL} - V_{FL}}{V_{NL}} \times 100\%$$

○ Example of Load Effect

Suppose a power supply is adjusted for a no-load output voltage V_{NL} of 10 volts, and its output drops to 8 volts under full load V_{FL}. Its load effect is as follows:

$$\% \text{ load effect} = \frac{10 \text{ V} - 8 \text{ V}}{10 \text{ V}} \times 100\% = 20\%$$

However, if the output drops to only 9.6 volts, the load effect is as follows:

$$\% \text{ load effect} = \frac{10 \text{ V} - 9.6 \text{ V}}{10 \text{ V}} \times 100\% = 4\%$$

As the examples show, quality increases as the percentage decreases. The load effect category includes any and all influences that the load has on the stability of output voltage. Typical load effect ratings can be lower than 0.01%.

Source Effect

Gradual and rapid (transient) line voltage changes can cause proportional variations in the output of an unregulated power supply. However, a regulator compares actual output with the desired output and adjusts circuit parameters in order to maintain the desired output voltage level.

Source effect, formerly called line regulation, describes the degree to which a desired output voltage can be maintained under input voltage variations. Values less than 0.01% are typical. Since there are reasonable limits in the ability of a regulator to accommodate input variation, an input range such as 115 volts rms ± 10% is usually specified.

Periodic and Random Deviation

Specifications called *periodic and random deviation* (PARD) and ripple factor describe the quality or smoothness of a DC voltage. The DC voltage can be measured with a VOM, and the AC ripple can be measured with an oscilloscope. Ripple factor is determined in this way:

$$\text{ripple factor} = \frac{\text{ripple}}{\text{output DC}}$$

$$\% \text{ ripple} = 100\% \times \text{ripple factor}$$

○ **Example of Calculating Ripple**

A supply with an output of 10 volts DC having 250 millivolts rms ripple would have the following percentage ripple:

$$\frac{0.250 \text{ V}}{10 \text{ V}} = 0.025$$

$$\% \text{ ripple} = 100\% \times 0.025 = 2.5\%$$

If improved filtering reduces the ripple to 10 millivolts rms, the percentage ripple is as follows:

$$\frac{0.010 \text{ V}}{10 \text{ V}} = 0.001$$

$$\% \text{ ripple} = 100\% \times 0.001 = 0.1\%$$

In general, PARD includes all periodic and random variation in the DC output voltage and is expressed in millivolts rms or millivolts pp. Typical values range from 1 millivolt rms in a 10-volt supply to less than 50 microvolts rms in a 10-volt supply.

Overload protection can include both fixed and adjustable current limiters, which protect a supply or a circuit under test from excessive current. Warm-up time and drift describe the long-term stability of a supply. For example, a specification may state that the output voltage will not drift more than 0.03% during an 8-hour period, following a 30-minute warm-up.

Output impedance describes how a supply appears as a source to a load and is typically expressed in milliohms and microhenrys. Transient recovery time describes the rate at which a supply can recover from sudden load changes, and it is a critical factor in digital systems. A quick-response supply can return the output to within 50 millivolts of the set value within 1 millisecond of the load transient. Digital readouts, remote control, overtemperature protection, and programming capabilities are some additional options available.

SUMMARY

For most electronic circuits, DC voltage is used as the power source. In addition, it is often used as a signal in the analog or digital representation of magnitudes. An ideal DC voltage has an accurately and precisely known magnitude, is perfectly smooth and stable, comes in a portable source, and has 0 ohms source impedance. The degree to which these ideal characteristics are attained indicates the quality of a DC voltage source.

The original and most common source of DC voltage is a battery. It can be a one-time primary cell or a rechargeable secondary cell. Batteries are generally categorized by voltage and active materials, which include zinc-carbon, lead-acid, lithium-iodine, and alkaline batteries.

Another DC voltage source is the laboratory power supply. A supply has four functional sections: transformer, rectifier, filter, and regulator. The input transformer isolates the supply from the source ground and changes the voltage level to the approximate value needed. A full-wave rectifier follows the transformer, and it changes the AC voltage to pulsating DC. A filter follows the rectifier, and it smooths the ripple and noise from the DC.

A power supply regulator is the final stage. It monitors the output voltage level and keeps it constant, even though the source voltage and output load may vary. While the other stages of a supply may have from one to four components, this stage has many components and is, by far, the most complex. The quality of this circuit's performance is directly related to the cost of a power supply.

A typical power supply has switches and fuses or circuit breakers at the input and the output. Voltage is adjusted with potentiometers and internal meters or with calibrated switches. Adjustable current limiters are also common and protect the supply from accidental overload. Most units have a handle and weigh less than 10 pounds. Although they can be moved from one bench to another, power supplies are not considered to be portable.

The most important specification of a power supply is output voltage, and it is usually indicated as a range, such as 0–10, 0–50, or 0–500 volts DC. The accuracy and resolution of the output voltage are other typical specifications. Regulation describes the impact of loading and can be calculated from no-load and full-load voltages. Low values indicate higher quality and can be less than 0.01% of the maximum output voltage. The load effect describes the degree to which changes in all loads affect the output voltage.

The impact of short-term and long-term line variations is indicated by the source effect. Since there are reasonable limits to the capability of a regulator to accommodate changes, an input range is also specified. Typical input values are 115 volts rms ±10%.

The quality or smoothness of the output voltage is described by the ripple factor. This factor is the proportion of AC rms ripple to DC output voltage and is expressed as a percentage. Typical values are less than 0.001, with the quality increasing as the value decreases. The periodic and random deviation (PARD) specification includes ripple along with all other periodic and random input variations. This specification is expressed in millivolts or microvolts, with typical values less than 1 millivolt rms for a 10-volt DC supply.

Additional specifications include output impedance, digital and analog metering, overload protection, warm-up time, drift, and weight. Also available are programming ability and remote control. The degree of quality specified and the options selected determine the cost of a laboratory power supply, which can range from hundreds to thousands of dollars.

BIBLIOGRAPHY

Batteries—a key ingredient for portability, Bob Margolin, Electronic Products, March 28, 1983.

Selecting power supplies for test, Donald R. Glancy, Test & Measurement World, June, 1983.

Qualifying a power supply, Victor Meeldijk, Electronic Products, January 11, 1984.

Understanding DC Power Supplies, Barry Davis, Prentice-Hall, 1981.

Fundamentals of Electronics, Douglas R. Malcolm, Jr., Breton Publishers, 1983.

QUESTIONS

1. Describe the characteristics of ideal DC voltage.
2. Name and describe three common sources of DC voltage.
3. Name four sections of a laboratory power supply, and describe the functions of each section.
4. Name four common controls of a DC power supply, and
describe the purpose of each control.
5. Name and describe five DC power supply specifications.
6. Define *source effect,* and explain why it is a concern.
7. Define *load effect,* and explain why it is a concern.
8. Compare half-wave, full-wave, and full-wave bridge circuit
characteristics.
9. What advantages does the full-wave rectifier have over the
half-wave rectifier? What advantages does the full-wave bridge rectifier have over the full-wave rectifier?
10. Why is ripple a problem in circuits?
11. Compare capacitor and inductor filters. Name an advantage and disadvantage of each type.
12. Name three purposes of a transformer in a power supply.
13. What is meant by PARD, and what does it include?
14. Describe how a series pass regulator responds to changes
in output voltage.
15. Describe the drift and warm-up time specifications and their
interrelationship.
16. What is a switching power supply?
17. What advantages does a switching power supply have over
a traditional rectifier-filter circuit?
18. What limitations does a series pass regulator have?

PROBLEMS

1. Determine the load effect of a power supply with an output
that drops from 29.0 to 25.0 volts DC when the load is applied.
2. Determine the load effect of a power supply with an output
that drops from 16.2 to 16.0 volts DC when the load is applied.

3. Determine the load effect of a power supply with an output that drops from 115.6 to 105.4 volts DC when the load is applied.

4. Determine the ripple factor of a power supply with an output voltage of 16 volts DC superimposed on 40 millivolts rms.

5. Determine the ripple factor of a power supply with an output of 280 volts DC superimposed on 70 millivolts rms.

6. Determine the ripple factor of a power supply with an output voltage of 105 volts DC superimposed on 0.15 volts rms.

7. Design a full-wave rectifier power supply that will provide an output of 12 volts DC with an input of 120 volts rms.

8. Determine the effect of adding a 10-microfarad capacitor and a 1000-ohm load to the circuit designed for Problem 7.

9. Design a full-wave bridge rectifier power supply that will provide an output of 18 volts DC with an input of 120 volts rms.

10. Determine the effect of adding a 5-microfarad capacitor and a 5000-ohm load to the circuit designed for Problem 9.

11. Design a shunt voltage regulator that will provide an output of 10 volts with an input of 36 volts, using a zener diode with $V_z = $ 15 volts and $I_{z\,max} = $ 75 milliamperes.

12. Design a shunt voltage regulator that will provide an output of 14 volts with an input of 24 volts, using a zener diode with $V_z = $ 18 volts and $I_{z\,max} = $ 75 milliamperes.

13. Locate five laboratory power supplies of similar voltage range in a catalog, and prepare a brief report comparing their quality specifications and their cost.

12

Potentiometers, Recorders, and Transducers

OBJECTIVES

This chapter introduces additional methods for the precise and accurate measurement of voltage. Instruments for both static and dynamic measurements are described, and some typical applications are presented.

> Describe the basic operation of a DC potentiometer.
> Explain how to measure a DC voltage with a potentiometer.
> Describe the operation and advantages of a servo-balancing measurement system.
> Describe the purpose and advantages of X–Y and strip chart recorders.
> Discuss the purpose of transducers.
> Name and describe three typical transducers.

POTENTIOMETERS

The familiar and useful adjustable resistor found in many electronic circuits is the potentiometer. Its ability to continuously divide voltage placed across it is the basic principle of operation of the measuring instrument called a *potentiometer*.

Basic Circuit

Potentiometers (the instruments) are used in laboratories for the accurate and precise measurement of DC voltage. They operate through the use of two methods described in earlier chapters, namely, null balancing and the comparison of unknowns with knowns.

The basic operation of a potentiometer is illustrated in Figure 12–1. A known voltage from an internal source is connected across a resistive potentiometer with a calibrated dial. This part of the circuit produces a continuously variable voltage of known magnitude that is connected to one side of a null detector. The unknown voltage is then connected to the other side. The potentiometer is adjusted until null is indicated, and the value of the unknown voltage is read from the calibrated dial. Of course, the actual potentiometer circuit is a bit more complicated.

A more complete, yet still simplistic, description of a potentiometer is presented in Figure 12–2. This diagram shows how some of the measurement concerns can be resolved. A decade divider accurately reduces the source voltage to smaller magnitudes so that the instrument does not totally rely on the linearity and resolution of the slide-wire. As in measuring voltage with an analog VOM, the intention is to utilize the largest amount of the measurement scale for greatest accuracy and resolution. The slide-wire is a smooth, linear potentiometer, and it provides a precise resolution exceeding one part per thousand. The null detector is similar to the one used in a bridge and includes a sensitivity switch. Calibration is provided

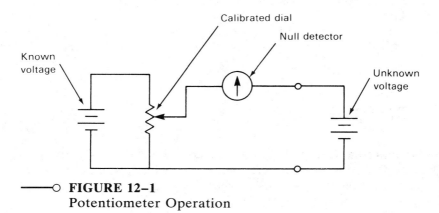

FIGURE 12–1
Potentiometer Operation

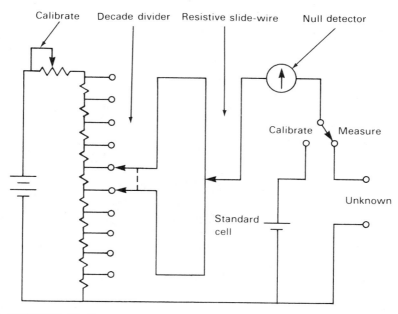

──○ **FIGURE 12–2**
Basic Slide-Wire Potentiometer

by a standard cell, which is described next. Circuit operation of the device will be described shortly.

Standard Cells

Figure 12–3 shows a typical unsaturated *standard cell,* which was the most common DC voltage standard for many years. Within the Bakelite case is an H-shaped, sealed glass vessel, as shown in Figure 12–4. Mercurous sulphate and mercury form the positive element, cadmium amalgam is the negative element, and a cadmium sulphate solution is the electrolyte. This combination produces a DC voltage between 1.01884 and 1.01964 volts, with the specific amount stamped on each cell before shipment. The voltage provided by a standard cell is typically within 0.005% of its stated value.

Standard cells require careful handling. Their current should not exceed 0.1 milliampere. In other words, you cannot measure their output with a VOM. In addition, they should not be exposed

──○ **FIGURE 12–3**
Standard Cell

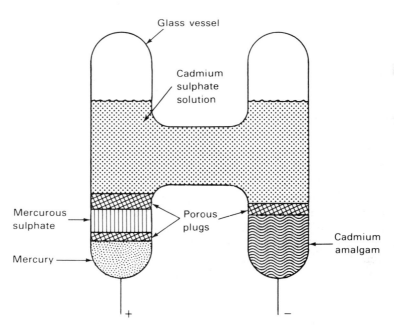

──○ **FIGURE 12–4**
Internal Parts of Standard Cell

to sudden temperature changes, and their temperature should not go below 4 or above 40 degrees Celsius. Finally, they should be handled gently and be recertified every two years.

DC Voltage Measurement

The measurement of voltage with a potentiometer is a simple process. The description here is somewhat general and presents the basic process used with most potentiometers.

Before a measurement is taken, the potentiometer must be calibrated. For calibration, the calibrate or standardize mode (see Figure 12–2) is used, and the slide-wire is placed on the value equal to the specified voltage of the standard cell. With some instruments, such as the one pictured in Figure 12–5, there is a calibrate position on the slide-wire and a separate control for the standard cell voltage. For these instruments, calibration is achieved by adjusting the calibrate potentiometer so that the voltage produced by the battery and divider and slide-wire source equals the voltage of the standard cell, as indicated on the standard cell control.

—○ **FIGURE 12–5**
Potentiometer

Once the potentiometer is calibrated, an unknown voltage can be measured. Measurement is accomplished by connecting the unknown voltage to the unknown terminals, switching to the measure mode, and adjusting the divider and slide-wire for a null indication. Null can be more sharply achieved by overriding the detector multiplier with the sensitivity switch once null is close. Then, the unknown voltage can be read from the indicated values of the divider and the slide-wire.

Some potentiometers utilize a series of cascaded decade dividers rather than one or two dividers and a slidewire. In a series of dividers, one can represent tenths of volts, one can represent hundredths of volts, and so on down to microvolt steps when, say, six dividers are used. Potentiometer accuracy depends on model and range but can extend from 0.03% to 0.0005%. The measurement range generally extends from 1.0 microvolt to 1.1 or 1.6 volts.

Volt and Shunt Boxes

The capabilities of a potentiometer can be extended with the addition of external options. A *volt box,* such as the one represented in Figure 12–6, is a voltage divider that accurately divides large voltages to give levels measurable with a potentiometer. A range switch directs larger voltages through more—and larger—series precision multiplier resistors. The reduced voltage that appears across the output resistor in the volt box series divider is then applied to the potentiometer input. This resistance also appears as a constant-resistance source to the potentiometer.

With a volt box, full-scale ranges up to 1500 volts can be reduced to 150 millivolts, well within the range of a potentiometer. Typical accuracy is 0.02%.

A *shunt box* extends the capabilities of a potentiometer to include current measurement. As Figure 12–7 shows, a shunt box consists of precision shunt resistors that are placed in series with the unknown current. This connection produces a voltage that can then be measured with a potentiometer. Typical ranges extend from 0.015 to 15 amperes, with an accuracy of 0.02%.

RECORDERS

Quite often, a single or occasional measurement is not enough to describe a parameter. Instead, continuous monitoring and automatic recording of its magnitude are required. In these cases, some sort

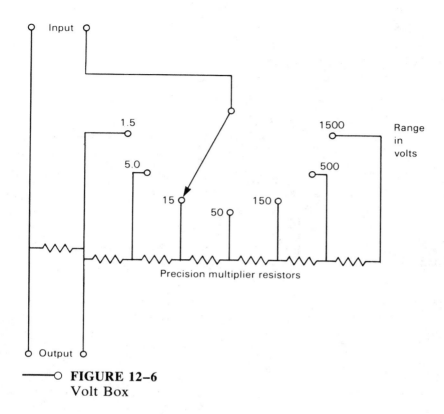

FIGURE 12–6
Volt Box

of *recorder* is used. Recorders are electromechanical instruments that produce permanent records of one or more parameters as a function of another parameter. Most often, the record is a roll of graph paper on which the recorder has automatically plotted the variations of one or more voltages as a funciton of time. Most recorders are designed around a servo-balancing system.

Servo-Balancing System

In previous chapters, we have examined two analog methods for determining parameter magnitude. In one method, such as found in a VOM, a pointer is moved upscale with energy derived from the source being measured. In the second method, such as found in a potentiometer, an operator adjusts a known source until it equals or gives a null balance with the unknown source.

　　Some recording systems use the first method with a D'Arsonval taut-band suspension pen. This device often has quick response but limited amplitude. The second method is generally preferred for re-

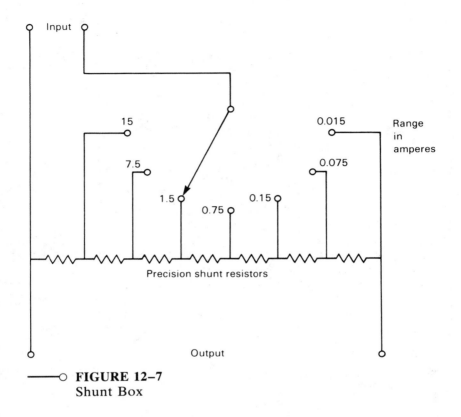

FIGURE 12–7
Shunt Box

cording devices because of its accuracy, sensitivity, and amplitude. It is accurate because it is designed around standards. It is sensitive because no energy is taken from the unknown source when null is achieved. *Servo-balancing systems* are null-balance systems driven by an amplifier and a motor.

As shown in Figure 12–8, the output of the servo-balancing system is a pointer deflection on a scale, which is also mechanically connected to a slide-wire potentiometer. Potentiometer position is directly proportional to pointer position, and both the potentiometer and the pointer are moved by the motor. The voltage to be measured is applied to the input of the servo amplifier, where the voltage from the slide-wire is subtracted from it. If they are unequal, an error voltage is produced. This error voltage is amplified and applied to the motor, which, in turn, moves the pointer and the slide-wire to the position where input and potentiometer voltages are equal.

If the slide-wire is excited with 1.0 volt, an input of 1.0 volt will cause full-scale deflection, while 0.5 volt will cause half-scale

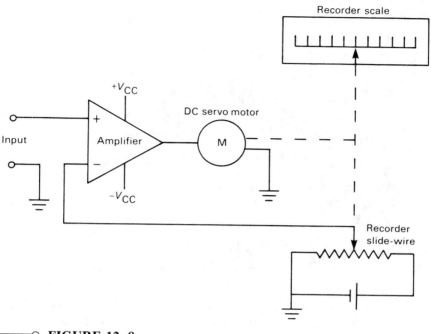

○ **FIGURE 12–8**
Servo-Balancing System

deflection. The servo system is polarity-sensitive, so it knows which direction to travel. It only moves when there is an error. Ranges can be changed by changing the slide-wire voltage, although input amplification-attenuation is a more common approach. Once again, since the servo system is a null-balance system, accuracy is primarily controlled by the slide-wire's voltage and linearity.

Strip Chart Recorders

Strip chart recorders provide a permanent record of the magnitude of a parameter as a function of time, and they do so automatically. There are two essential parts to the typical strip chart recorder: the pen deflection system and the chart drive system. Pen position is controlled with a system similar to the servo system. While most recorders have only a single channel—that is, have one pen—many channels are possible, as in the eight-channel recorder shown in Figure 12–9. The chart is generally driven by a synchronous motor through changeable gears.

○─── FIGURE 12–9
Eight-Channel Strip Chart Recorder (Photo courtesy of
Watanabe Instruments Corporation)

There are many specifications to consider in the selection of a
chart recorder. The following list provides an indication of the range
and choices in options:

Number of channels available, from 1 to 12;
Input impedance, usually 1 megohm;
Writing method, ink or hot stylus;
Chart width, from 2 3/4 to 10 inches;
Chart speed, from 1 millimeter per hour to 50 millimeters per
    second;
Drive, continuous or stepped;
Accuracy, from 0.25% to 1.0%;
Linearity, typically 0.15%;
Resolution, typically 0.1%.

The frequency response of a chart recorder varies greatly and
depends on the amplitude of pen displacement. For a full-scale de-

flection of 10 inches, a 1-hertz bandwidth is typical. Yet, some recorders can plot small-amplitude signals up to 120 hertz. The
capabilities of a system will be specified with an amplitude-band-
width factor.

With the addition of input options, it is possible to continuously
measure a wide range of AC and DC voltages, frequency, and many
other parameters. Printers for time, date. and individual events can
often be added, too.

Strip chart recorders find extensive use in situations where low-
frequency phenomena occur and permanent records are needed. For
example, an AM radio station might maintain a 24-hour continuous
record of final output voltage, antenna current, frequency deviation,
and antenna phase with a multichannel, low-speed chart recorder.
In this application, the only additional components needed are transducers to convert the parameter of interest to voltages within the
input range of the recorder.

Another application of strip chart recorders is the measurement
of temperature as a function of time. A multi-input recorder could
plot the control voltage of an environmental chamber along with the
output voltage of thermocouples placed in selected places within the
chamber. This system would provide both an immediate indication
and a permanent record of the response time and the temperature
distribution of the chamber.

X–Y Recorders

Strip chart recorders, as just described, plot one or more parameters
as a function of time. An *X–Y recorder* plots one parameter as a
function of another. Figure 12–10 shows a typical X–Y recorder.
Rather than use long rolls of paper, it uses single sheets. While its
function can also be obtained with an oscilloscope, the X–Y recorder
or plotter provides a permanent record, a larger display, and greater
accuracy. However, it does not have the frequency response of an
oscilloscope.

The typical X–Y recorder utilizes two independent servo-balancing systems. One system, for the X (horizontal) channel, moves
a carriage from left to right. The other system, for the Y (vertical)
channel, moves the pen up and down on the carriage. As with most
other recorders, both inputs have switchable ranges. Zero can be
located at one end or the other or any place in between.

The X–Y recorder is generally used to indicate and record the
relationship between independent and dependent variables. For example, it might be used to record the relationship between stress

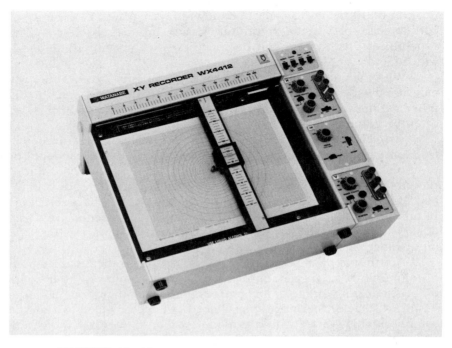

—○ **FIGURE 12–10**
X–Y Recorder (Photo courtesy of Watanabe Instruments
Corporation)

and strain of a physical structure. The force applied to the structure—the stress—can be expressed as a voltage by use of transducers
(which are described shortly). This voltage becomes the X, or independent, variable input of the recorder. The amount of elongation—the strain—can also be measured with a transducer and
expressed as a voltage. This voltage becomes the Y, or dependent,
variable input of the recorder. The X–Y recorders are appropriate
for events like stress and strain that take minutes or hours. Events
occurring in milliseconds or less still require oscilloscopes.

Some added features available for X–Y recorders include dual-
Y input, automatic feed for single sheets, and roll paper. Both chart
and X–Y recorders often offer retransmitting potentiometers that
are mechanically attached to the essential slide-wire potentiometer.
Exciting this potentiometer with a separate voltage provides an analog voltage for the operation of a parallel measuring instrument.

TRANSDUCERS

Transducers are often used to convert one physical dimension to electric signals for measurement, monitoring, and control. The magnitude of sound, temperature, light, motion, and many other phenomena can be represented by voltage, resistance, inductance, or capacitance. With electronic circuits, this information can then be electrically displayed, measured, and recorded or processed by a computer. Some typical transducers are thermocouples, strain gages, and linear, variable differential transformers.

Thermocouple

A *thermocouple* is a transducer used in the measurement of temperature. When dissimilar metals come in contact with one another, minute voltages are produced. A thermocouple is a pair of unlike conductors, bonded to form a junction at one point. The magnitude of the voltage it produces depends on the combination of metals and the temperature it is measuring, but the voltage is predictable and linear. Generally, it is in the microvolt-per-degree range.

The bonding of the junction where the two metals meet is accomplished by pressure or heat. Soldering is not used because soldering would introduce additional metals, tin and lead. Thermocouple metal combinations often used include iron-constantan, copper-constantan, nickel-iron, and platinum-rhodium/platinum. Iron-constantan thermocopules have a usable range from −300 to +1600 degrees Fahrenheit. Platinium-rhodium/platinum thermocouples range from 1600 to 3100 degrees Fahrenheit.

A typical thermocouple circuit for temperature measurement is shown in Figure 12–11. One thermocouple cannot be used alone because two additional and unlike junctions would be produced where it connects with the measurement instrument. Instead, two thermocouples are used.

The combination shown in Figure 12–11 works as follows. The temperature being measured produces a voltage at the measurement junction. Another voltage is produced by the reference junction, which is generally at room temperature in an insulated box; its temperature is measured with a thermometer. While two additional voltages are produced at the contact points of metal A and the potentiometer, they are of equal and opposite magnitude and, therefore, cancel. These canceling voltages are the reason the additional section of metal A is used.

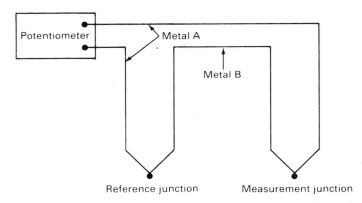

──○ **FIGURE 12–11**
Thermocouple Circuit

Temperature is determined by measuring the voltage with the potentiometer, reading its equivalent temperature from a thermocouple chart, and adding room temperature, which has been electrically subtracted by the reverse-connected reference junction. Thermocouple manufacturers provide users with temperature-emf tables so that temperature can be measured with thermocouples and a potentiometer. Iron-constantan thermocouples come in ranges of −310 to −100 degrees Fahrenheit, −100 to +600 degrees Fahrenheit, 600 to 800 degrees Fahrenheit, and 800 to 1400 degrees Fahrenheit, with a normal accuracy of ±4 degrees.

Strain Gage

A *strain gage* is a resistive strip that changes resistance as it is elongated (strained). It is used in the measurement of displacement. Figure 12–12 indicates the relative shape and size of typical strain gages. Usually, gages are physically attached to the material to be measured and electrically connected as part of a bridge circuit, as shown in Figure 12–13. Temperature, undesired motion, or other undesired variables are compensated for by additional strain gages placed in physical and electrical positions to produce canceling effects.

When the strain gage bridge of Figure 12–13 is in operation, the lower left gage (the measurement gage) elongates as the object of measurement extends in the direction of the arrow. The amount of extension is determined by obtaining null with the balance poten-

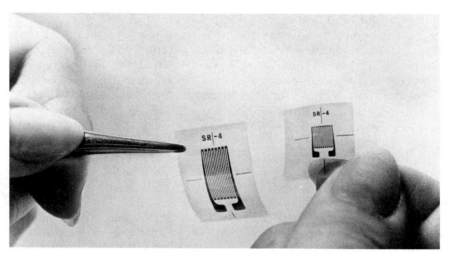

───○ **FIGURE 12–12**
Strain Gages (Photo courtesy of BLH Electronics/Division of Bofors Electronics, Inc.)

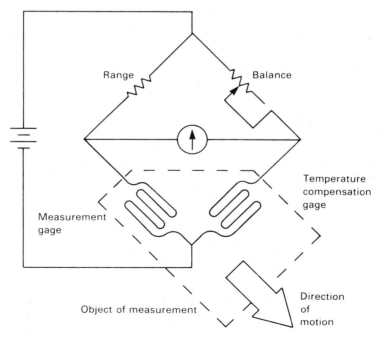

───○ **FIGURE 12–13**
Strain Gage Bridge

tiometer and reading the indication. Quite often, the elongation to be measured occurs as a result of temperature variation rather than force. Error is compensated for by a temperature compensation gage, as shown in Figure 12–13. It is physically located so that it does not elongate when motion occurs in the direction of interest. It does sense the same temperature because it is also physically attached to the object. Because of its electrical placement in the bridge, its resistance changes as a function of heat cancel the changes of the measurement gage.

For continuous or dynamic measurement, the balance arm of the bridge is a slide-wire potentiometer of a servo-balancing recorder. Error from power source drift is eliminated by exciting the strain gage and the slide-wire from the same source. Once the system is assembled, the recorder can be zeroed left, right, or center, depending on the motion expected. As usual, this system must first be calibrated; calibration is done by applying a known deflection and recording the response.

LVDT

A *linear, variable differential transformer* (LVDT) is another common displacement transducer. This device is a cylindrical transformer with a core that moves freely through the transformer center.

As Figure 12–14 shows, the LVDT has a primary and two series-connected secondaries. The secondaries are connected in opposition so that the output is of one phase with the core at one end, zero in the middle, and of the opposite phase at the other end. That

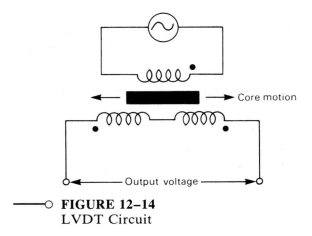

───○ **FIGURE 12–14**
LVDT Circuit

is, the voltage increases from zero to maximum of one phase as it extends from the center position to one end. It increases in a similar manner from zero to maximum as it extends from the center position to the opposite end. However, it has a 0-degree phase shift on one side of center and a 180-degree phase shift on the other side. As shown in Figure 12–15, the output voltage is proportional with the core's distance from the center, and the phase will be 0 or 180 degrees, depending on the core's direction from the center.

The overall range of an LVDT can vary from 0.005 to 10 inches, depending on the needs of the application. They can sense changes in microinches, and they have a linearity of around 0.1%. In a typical measurement situation, the transformer body is secured to one point, and the core is secured to another. Before displacement is measured, zeroing must be done. Zeroing can be done mechanically, but it is usually done with the adjustment of another arm of the bridge. The output voltage is used as one input to a servo system, with a recorder slide-wire as the other input. Both are excited from the same AC

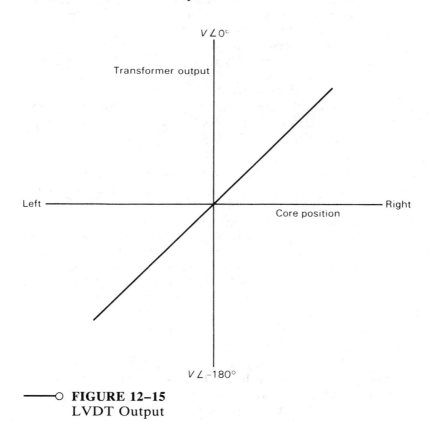

──○ **FIGURE 12–15**
LVDT Output

source. Since this device is a transformer, it is sensitive to changes in frequency and temperature.

The most common application for an LVDT is the measurement of elongation. Generally, it is attached to the object to be measured through a system of gears or levers. These mechanisms reduce or increase the displacement to a magnitude that is within the range of the LVDT. Clamps and positioning pins secure the transducer to two positions, and the distance between these two positions is called the gage length. The change in the gage length is sensed when the coil is secured to one position and the core to the other.

Additional Transducers

The use of transducers extends far beyond the electronics industry. Industrial manufacturers use pressure and flow transducers for the monitoring and control of processes. Hospitals use transducers for the monitoring of a patient's pulse, respiration, and temperature. Aircraft utilize transducers to measure and monitor acceleration, altitude, speed, attitude, temperature, and so forth.

With the aid of resistive, inductive, capacitive, piezoelectric, and semiconductor transducers, static and dynamic physical phenomena can be accurately represented by V, I, or R analogs for processing by electronic instrumentation. Any component or material whose electrical characteristic changes as a function of another parameter can be, and probably is, used as a transducer. Some typical examples are given in the following listing:

Transducer	Parameter Measured
Piezoelectric crystal	Pressure
Photovoltaic cell	Light
Reluctance pickup	Vibration
Thermistor	Temperature
Photo diode	Light
Tachometer	Angular velocity

SUMMARY

A potentiometer is an instrument designed to provide accurate and precise measurements of DC voltage. Built around a system of dividers and a slide-wire, the potentiometer measures voltage by the methods of null balancing and comparison of unknowns with knowns.

Quite often, an internal standard cell is used for calibrating the instrument. The typical unsaturated standard cell produces a stable, known voltage between 1.01884 and 1.01964 volts. It operates through the interaction of mercurous sulphate and mercury on the positive side and cadmium sulphate and cadmium amalgam on the negative side.

For DC voltage measurement, the unknown voltage is connected to the input terminals of the potentiometer, the slide-wire and divider are adjusted for null balance, and the value is read from the slide-wire and the multiplier-divider dial. Typical ranges extend from 1 microvolt up to 1.1 or 1.6 volts. Accuracy depends on model and range but may extend from 0.03% to 0.0005%. Higher voltages or current can be measured with a potentiometer when a volt box or a shunt box is added to the circuit.

Recorders are used to provide continuous monitoring and a permanent record of parameter values. The heart of many recorders is a servo-balancing system, which automatically adjusts a recording pen to a position analogous to parameter magnitude. It does so by comparing the two variables, detecting error, and driving a servo motor to correct the error.

Strip chart recorders record parameter value, for up to as many as 12 variables, as a function of time. The variable need only be presented as a voltage at a magnitude compatible with the recording system. Recorder input adapters are available for the accommodation of a wide range of variables. The typical recorder has a 10-inch wide chart whose travel speed can vary from 1 millimeter per hour to 50 millimeters per second. Input impedance is 1 megohm, accuracy is from 0.25% to 1.0%, linearity is 0.15%, and resolution is 0.1%.

The X–Y recorders utilize two servo-balancing systems, one to move a carriage in the horizontal (X) direction and the other to move a pen vertically (Y) along the carriage. This device plots one parameter as a function of another. While it produces a permanent, larger, and more accurate record than that of an oscilloscope, it does not have the frequency response of an oscilloscope.

Transducers are used to convert one parameter to another. In measurement instruments, transducers convert temperature, pressure, motion, or some other physical phenomenon to an electric signal, such as voltage, current, or resistance. Typical transducers are thermocouples, strain gages, and LVDTs. They are used in a wide range of situations including medical electronics, avionics, and manufacturing process control.

Thermocouples are used to measure temperatures both well above and well below freezing. They do so by producing a voltage proportional to temperature. The voltage is then measured with a potentiometer and converted to temperature with the aid of tables.

Strain gages and LVDTs are used to measure elongation. The resistance of a gage changes as a function of elongation. Strain gages are typically bonded to a surface in permanent installations. Then, resistance changes are measured by balancing the bridge, after it has been calibrated with a standard. The LVDTs are usually part of mechanisms that are attached temporarily. They require AC, while the strain gage can be used with AC or DC. Voltage change as a function of displacement is sensed, measured, and interpreted with calibration tables or a servo system.

BIBLIOGRAPHY

Sensors, signal conditioning, and systems, Bob Butler, Electronic Products, March 4, 1983.

Digital signal processing extends range of waveform recorder, Control Engineering, April, 1983.

Signal recorder analyzes, displays, and plots signals, D. Bursky, Electronic Design, December 8, 1983.

Industrial Electronics, James T. Humphries and Leslie P. Sheets, Breton Publishers, 1983.

Optical Transducers & Techniques in Engineering Measurement, A.R. Luxmoore, ed., Applied Science Publishers, 1983.

QUESTIONS

1. Describe the basic circuit and operation of a DC potentiometer.

2. Compare the accuracy and range of a DC potentiometer with those same features in a nonelectronic VOM.

3. Describe two applications that you might have for a DC potentiometer in a laboratory.

4. Compare the use and care of a standard cell with the use and care of a typical primary or secondary cell.

5. Explain the purposes of volt boxes and shunt boxes.

6. Describe the operation of a servo-balancing system.

7. Describe two situations where servo-balancing systems are used.

8. State two advantages of servo-balancing systems.
9. Define *transducer*.
10. Explain why transducers are used.
11. Name two transducers, and describe how they are used.
12. Explain the purpose of a reference junction in a thermocouple.
13. What advantage does a strain gage have over an LVDT?
14. Describe the operation of a chart recorder.
15. Explain two advantages of a chart recorder.
16. What advantage does an X–Y recorder have over an oscilloscope? What is its disadvantage?
17. Describe the operation of an X–Y recorder.

13

Analyzers

OBJECTIVES

This chapter introduces spectrum and network analyzers. Spectrum analyzers combine features of sweep generators, counters, and oscilloscopes to produce a versatile measurement system. Network analyzers have similar features plus the features of computer control, monitoring, and analysis.

Name the sections of a spectrum analyzer, and describe the function of each.

Name four specifications of a spectrum analyzer, and describe typical values.

Describe an application of a spectrum analyzer.

Name the sections of a network analyzer, and describe the function of each.

Name two specifications of a network analyzer, and describe typical values.

Describe an application of a network analyzer.

SPECTRUM ANALYZER CIRCUITS

A *spectrum analyzer* can be described as a frequency-selective voltmeter. It consists of an input-ranging circuit, a system for passing the desired frequency and blocking all others, and a voltage-indicating device. The latter device is generally a CRT. Most models vary in the manner in which frequencies are passed and blocked.

Purpose of the System

A common measurement by a technician is the frequency response of a device or system such as an audio amplifier. The process is simple but time-consuming. A variable-frequency audio oscillator is connected to the amplifier input, and a voltmeter is connected across the load. The input frequency is varied, and the operator ensures that the input voltage remains constant. Output voltage as a function of frequency is measured and recorded. Gain as a function of frequency is then calculated and plotted, producing the typical frequency response curve, such the curve shown in Figure 4-6.

The purpose of a spectrum analyzer is to provide the input signal for the frequency response evaluation of a device (the test just described) and then to display the device's output as a function of frequency. A spectrum analyzer provides the input signal and sweeps that signal over a preselected range. Another signal provided by the analyzer is a DC voltage that is proportional to the output frequency. This signal is used to sweep the display CRT horizontally. The output of the test circuit is the vertical input of the CRT. A frequency response curve for output voltage versus frequency is then displayed on the CRT.

There is one major limitation to the use of the spectrum analyzer as just described. That approach assumes that the output frequency is the same as the input frequency. That is, it assumes that when the input frequency is 1 kilohertz, the output voltage is comprised of only 1-kilohertz energy. The assumptions that no harmonics were generated and that there was no distortion are usually not allowed.

However, as we know, a 120-volt, 60-hertz power transformer has an output of the fundamental frequency and the third harmonic. That is, the desired voltage is at 60 hertz and an additional amount is at 180 hertz. Measurement of the output voltage with a VOM would indicate neither the presence nor the amount of harmonic content. In the case described earlier, the input magnitude was kept constant, the input frequency was swept from low to high, and the output magnitude at the input frequency was measured. So what is needed here is a system in which the input magnitude and frequency are kept constant and the output magnitude is measured at multiples, or harmonics, of the input frequency.

The harmonic content of a transformer can be determined by passing the output voltage through a bandpass filter before its magnitude is measured. If the filter has a narrow bandpass of around 60 hertz, the fundamental-frequency output voltage can be measured. Changing the filter so that it has a center frequency of 180 hertz allows the third-harmonic content to be measured.

Providing the sweep frequency input signal is just one task of the analyzer. (This sweep frequency is produced with a voltage-controlled oscillator, described in Chapter 9.) Another task—and a critical one—is simultaneously sweeping the pass frequency input of the measuring circuitry. Close tracking is needed between the two tasks to ensure that the frequency being viewed is the one going in. The width of frequency viewing cannot be infinite since harmonics will pass. Yet, the frequency width cannot be zero, for then nothing will pass. The following subsections describe three methods of resolving this dilemma: a sweep frequency system, a multiple-filter system, and a time comparison system.

Sweep Frequency System

Figure 13–1 shows the block diagram of a traditional spectrum analyzer. We will consider it when it is used to measure the characteristics of a power transformer. Assume that 120 volts at 60 hertz are applied to the transformer primary. The transformer secondary is

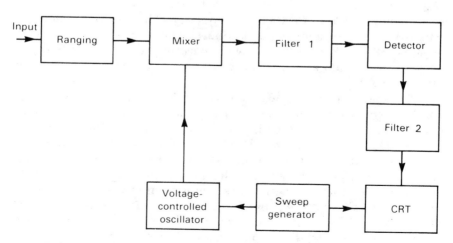

───○ **FIGURE 13–1**
Sweep Frequency Spectrum Analyzer

then connected to the input of the analyzer. Our discussion of circuit operation will begin at the sweep generator.

The sweep generator produces a sawtooth voltage ramp that rises at a constant rate to a positive peak, drops quickly, and begins the cycle again. This ramp, shown in Figure 13–2A, serves two purposes. First, the ramp is connected to the horizontal input of a CRT display system to sweep the beam from left to right at a constant rate and then recycle. Second, the ramp is connected to a voltage-controlled oscillator (VCO) to sweep the VCO output frequency over a predetermined range. The VCO output (Figure 13–2C) is then sent on to the mixer.

Meanwhile, the signal to be evaluated (Figure 13–2B) is applied to the ranging circuit. The ranging circuit is an amplifier-attenuator used to ensure that the incoming voltage is of workable magnitude. When the signal being measured is of proper amplitude, it is also sent to the mixer. A mixer is a circuit designed to receive two inputs and produce one output. The output of the mixer consists of two frequencies, the sum and the difference of the two input signals.

Filter 1 of Figure 13–1 is a low-pass filter. It removes the sum signal and passes the difference signal. The detector removes the negative half of the mixed signal, changing the signal to varying DC.

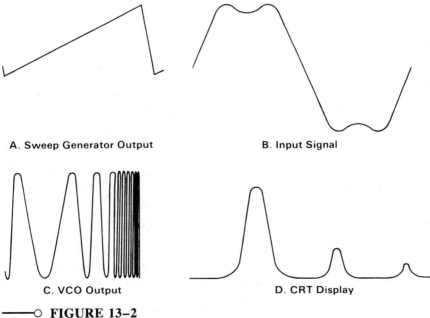

A. Sweep Generator Output   B. Input Signal

C. VCO Output   D. CRT Display

——○ **FIGURE 13–2**
Waveforms for Sweep Frequency Spectrum Analyzer

Filter 2 integrates, or smooths, the detected signal and sends it to the CRT display. Thus, the oscillator, mixer, filter, detector, and filter system operates much like a superheterodyne receiver.

Figure 13–2D shows the output waveform that is displayed in the measurement just described. The horizontal axis represents frequency, as controlled by the sweep generator. The vertical axis represents the component of transformer output voltage as a function of frequency. As the figure shows, this transformer has the expected output at 60 hertz, a small amount of third-harmonic voltage (180 hertz), and an even smaller amount of fifth-harmonic voltage (300 hertz).

If a traditional measurement had been made instead of the analyzer measurement, a voltmeter would have indicated the average or effective voltage level, and a counter would have indicated only the fundamental frequency. Thus, the spectrum analyzer has important advantages, but it also has limitations. It is slow, often taking seconds per sweep, and it has a limited range.

Multiple-Filter System

The multiple-filter system in Figure 13–3 is used to analyze signals up to a few hundred kilohertz. Like the operation of the sweep frequency system, its operation begins with a ramp-producing sweep generator. Unlike the sweep frequency system, though, it is fast. The ramp is used to drive the horizontal sweep of the CRT display and to operate a high-speed scanner. Signals to be analyzed are applied to the inputs of a bank of parallel-connected, narrow-band filters. Each filter will pass a narrow frequency band, and adjacent filters have a slight frequency overlap, as shown in Figure 13–4.

In the analysis process, the scanner sequentially transfers the output of each filter to the detector, the filter, and the vertical input of the CRT display. At the same time, the sweep generator correlates the horizontal position to produce a display of voltage amplitude versus frequency band.

Some error can occur with this system because none of the filters have the ideal, desired, square-shaped frequency response curve. The pass–no-pass limits are not specific frequencies, so the sharp yet bell-shaped bandpass can produce some attenuation away from each filter's center frequency. If the center frequencies of all the filters are not equally spaced, the attenuation can be inconsistent, causing even more error. However, these limitations do not affect this system's wide use in audio frequency signal analysis.

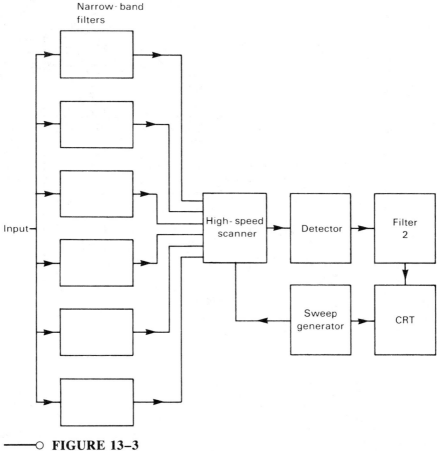

⎯⎯◯ **FIGURE 13–3**
Multiple-Filter Spectrum Analyzer

Time Compression System

The analysis process and resulting display waveforms described in
the previous subsections involved the frequency domain, that is, amplitude as a function of frequency. We will now consider the time
domain, that is, amplitude as a function of time. A time domain is
the way we typically view a sine wave or other signal on an oscilloscope.

In the time compression system, shown in Figure 13–5, an incoming signal is rapidly scanned as a function of time. Magnitudes

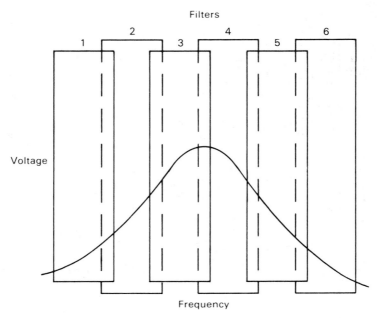

—o **FIGURE 13–4**
Multiple-Filter System Bandpass

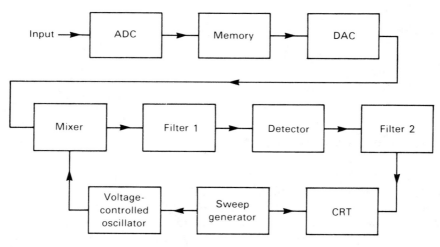

—o **FIGURE 13–5**
Time Compression Spectrum Analyzer

are then converted from analog to digital form in the ADC and stored in memory. Additional scans and conversions occur, and the resulting data is sent to memory. This process allows many rapid passes over a given waveform, ensuring that one-time characteristics are given appropriate consideration.

When the predetermined number of passes have been completed, the data is read from memory and sent to a digital-to-analog converter (DAC). An analog summary of the input waveform is constructed and sent to circuitry similar to the sweep frequency system described earlier. The final display rate is much faster than the rate of the initial sweep since many passes, typically in the hundreds, are made in data acquisition.

SPECTRUM ANALYZER SPECIFICATIONS

A typical spectrum analyzer is pictured in Figure 13–6. The specifications for the analyzer as described by the manufacturers vary greatly according to type of system and the range of added capabilities. There are, however, specifications that are common to most systems. Five basic specifications are described here.

Frequency Range and Accuracy

The frequency range that spectrum analyzers can process is divided into moderate-width bands. Audio range systems operate from frequencies as low as 0.01 hertz up to 300 kilohertz. Radio frequency systems span a range from 100 kilohertz to 1200 megahertz. Microwave systems extend beyond 1200 gigahertz. A low-to-high frequency ratio of 10^6 is common.

Frequency resolution depends on frequency range and can be under 1 hertz with an AF spectrum analyzer. Microwave analyzers can view a sector as narrow as 100 hertz.

The accuracy of the frequency between two points on the display is a common specification and is typically between 2% and 10%. In more expensive microprocessor-controlled systems, center-frequency is at least within -0 to $+0.2\%$ of the frequency span setting. Long-term drift is less than 1×10^{-7} per month of frequency.

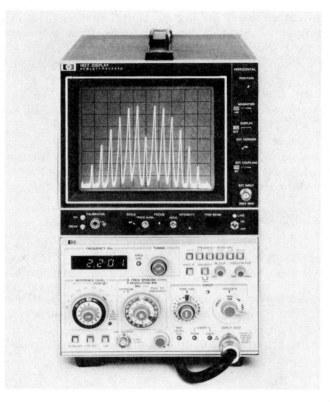

———○ FIGURE 13–6
Spectrum Analyzer (Photo courtesy of Hewlett-Packard
Company)

Amplitude Range and Accuracy

The typical spectrum analyzer used in a laboratory has an input
range described in terms of decibel-milliwatts. A 50-to-70-decibel-
milliwatt range is typical, such as a range from −60 to +10 decibel-
milliwatts. Some systems have an input range from −117 to +30
decibel-milliwatts. An input impedance of 50 ohms is typical.

Amplitude accuracy is described in terms of flatness of frequency response and is typically ±1 decibel. In addition to having
relative amplitude accuracy, more expensive systems have amplitude
calibrators and can determine absolute amplitude accuracy to better
than ±0.2 decibel.

SPECTRUM ANALYZER APPLICATIONS

Spectrum analyzers are used for a wide range of measurements. Signal purity, modulation, component linearity, pulse shape analysis, and noise measurement are just a few applications. Determining the quality of a filter and the characteristics of an oscillator will be described here. While systems vary in operation, features, and capabilities, the system described here is representative of typical systems.

Filter Quality Measurement

In this discussion, an intermediate frequency (IF) filter is evaluated for Q and bandwidth with a spectrum analyzer. The filter's ability to pass the desired center frequency and reject adjacent frequencies is also investigated. Our system consists of an RF section, an IF section, a display, a tracking generator, and a counter, as shown in Figure 13–7. The output of the filter enters the system at the RF input connector. The input of the filter must be a broadband, externally generated signal.

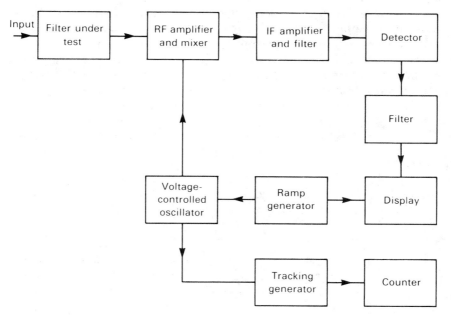

──○ **FIGURE 13–7**
System for Measuring Filter Quality

Before the spectrum analyzer is turned on, many adjustments must be made. The center frequency control must be set; in this case, we use 455 kilohertz. Scan width is adjusted next. Since adjacent-channel rejection is the concern, 50 kilohertz is reasonable. The input attenuator setting depends on the magnitude of the incoming signal. As always, we begin with maximum attenuation.

Scan time can be adjusted for convenience when the observation process begins. The display needs the usual adjustments for intensity and focus. With the filter properly powered and connected and the basic adjustments made, the observation process can continue. So, we next adjust the scan width, attenuator, and scan time controls until a waveform similar to the one shown in Figure 13–8 is obtained.

The tracking generator produces a continuous wave (CW) carrier with the same frequency as the frequency of the RF sweep generator. One application of this combination is to produce a small vertical mark on the display, as is also shown in Figure 13–8. When used as a marker generator, the tracking generator moves the mark horizontally to align with a predetermined frequency. This frequency is then displayed on the counter. The log/linear level control adjusts the vertical amplitude reference points and is calibrated in decibel-milliwatts or volts.

We now have two approaches for observation. We can simply view the displayed waveform to observe the linearity of filter response. Or we can move the mark to specific points of interest, such as maximum gain and −3 decibels, and read the frequency of each point from the counter. In addition, points of concern can be detected and identified.

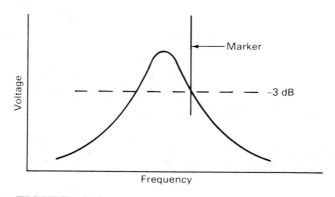

——○ **FIGURE 13–8**
Output Waveform for IF Filter Frequency Domain

Determination of Oscillator Characteristics

An oscillator is used to produce a sine-shaped voltage of a specific frequency. Its output can be measured with instruments described earlier in this text. For example, we can measure the rms voltage with a meter, the peak-to-peak voltage with an oscilloscope, and the fundamental frequency with a counter. However, an oscillator could have other characteristics, such as distortion, harmonics, and parasitic oscillations, to name just a few. This section discusses methods for determining such characteristics with the spectrum analyzer.

For this test, the signal to be evaluated is connected to the analyzer input. The design output frequency is used as the analyzer center frequency, and a scan width is selected. The scan width should be of a size that includes the fourth harmonic. Once again, voltage as a function of frequency can be detected and quantified. The value of an analyzer is again demonstrated in this test.

Figure 13–9A shows an oscillator waveform we might see on an oscilloscope. It is in the time domain, displaying voltage as a function of time. For this waveform, we can measure peak-to-peak voltage and period. We can also observe that the waveform is distorted and has what appears to be noise.

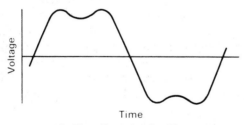

A. Time Domain, Oscilloscope

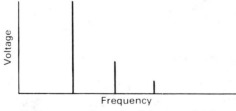

B. Frequency Domain, Spectrum Analyzer

─────○ **FIGURE 13–9**
Oscillator Output

Figure 13–9B shows how a spectrum analyzer would display the same signal in the frequency domain—that is, voltage as a function of frequency. The frequency domain signal shows a large voltage at 10 kilohertz, the desired fundamental frequency. A smaller, but obviously present, voltage exists at 20 kilohertz, the second harmonic. This harmonic is probably the cause of the waveform's distortion. Further to the right is an even smaller voltage. If we move the marker to that point, the counter identifies this frequency as 1.2 megahertz. It is a parasitic oscillation, and it is unwanted.

This test has achieved much. It indicated the presence of distortion and showed that it was the result of second-harmonic content. It indicated the presence of a parasitic oscillation and its frequency. Knowing this much about the undesired characteristics would help greatly in determining how to eliminate them.

Other Measurements

The list of spectrum analyzer applications is quite long. Applications include detecting the spectral purity of a CW signal by looking for undesired sidebands. Determining the modulation percentage of an AM signal is another common analyzer application. Also, the carrier and modulation frequencies of an FM signal can be viewed. And low-level distortion not visible in the time domain can be detected in the frequency domain. Finally, signal-to-noise ratios can be measured.

Interfacing a spectrum analyzer with a computer provides greater versatility in automatic testing of systems. The rapid, automatic measurement of frequency response characteristics can quantify many individual components within a system. Of course, as spectrum analyzers become computer-controlled, their already high price gets higher. Yet, their cost effectiveness also improves. They provide a lot of evaluative data in a very short time. So, their place in electronics can only increase.

NETWORK ANALYZERS

As just described, a spectrum analyzer is a frequency-selective voltmeter. A *network analyzer* is a spectrum analyzer with the added capability of exciting the system or component to be evaluated and analyzing the output as a function of the input. The transfer function of a component or system is then presented in terms of gain and phase versus frequency.

System Operation

The operation of the network analyzer system can be best described through two common applications, determining the impedance of a passive component and the frequency response of an amplifier. We will consider the impedance measurement first.

Figure 13–10 shows the connections for component impedance analysis. The component is excited by a sweep oscillator, which, in turn, is controlled by the analyzer.

During the measurement process, a constant voltage is applied across the component under investigation. This voltage is also applied to the reference input of the analyzer. With a current transformer, the component's current can be monitored and applied to the signal input of the analyzer. The quotient of these two inputs will be used to calculate the impedance of the component.

A microprocessor-based internal controller rapidly sweeps the oscillator frequency over a preprogrammed range. The two signals entering the analyzer are amplified and attenuated to workable size. Samples of each signal are taken at a rate that is programmed according to the complexity of the signals. Low-distortion signals may require as few as 16 samples per cycle, while high-distortion signals may require up to 256 samples per cycle.

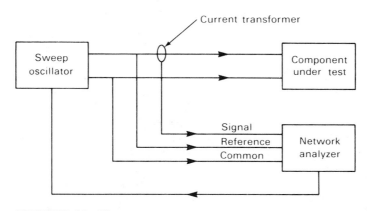

—○ **FIGURE 13–10**
Impedance Measurement with Network Analyzer

Samples are converted from analog to digital data by successive approximation and are stored in memory. Upon completion of the programmed number of sweeps, the requested data are calculated. In this case, the data are impedance magnitude and phase angle as a function of frequency. Depending on the analyzer model, data can appear on an LED or LCD display, a CRT, or a printer. A summary printout might look like the table that follows:

Frequency (Hertz)	Impedance (Ohms)	Phase Angle (Degrees)
200	500	−87
300	450	−85
400	390	−80
500	330	−70
600	250	−60
700	180	−45
800	140	−32
900	110	−25
1000	100	0
1200	100	+25
1300	160	+36
1400	210	+46
1500	240	+54
1700	300	+62
2000	360	+70
2500	420	+79
3000	460	+82

These data indicate that the circuit has a fundamental bandpass frequency, minimum opposition, and no phase shift at 1 kilohertz. Below that frequency, there is increasing opposition and phase lag. Above the fundamental frequency, there is increasing opposition and phase lead.

Now, we will consider the frequency response measurement. Figure 13–11 shows a configuration for measuring amplifier gain and frequency response with a network analyzer. Once again, an oscillator provides an input signal of constant voltage and varying frequency. Also as in the previous case, the analyzer monitors and compares two signals. In this case, however, one signal is the amplifier input and the other is the amplifier output. The quotient of these two signals is amplifier gain as a function of frequency, or, as it is more commonly called, frequency response.

Figure 13–11 also shows another feature and benefit of most network analyzers, the IEEE–488 or RS–232 data bus. The data bus allows communication between a controller, a computer, a display,

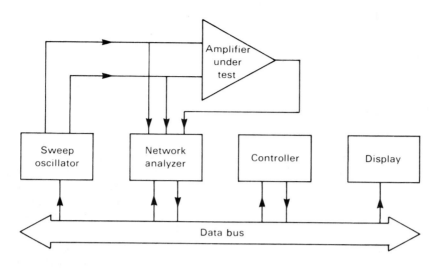

○ **FIGURE 13–11**
Frequency and Gain Measurement with Network Analyzer

a graphic printer, or some other external device. In automatic test systems, data can be externally developed, stored, and initiated as required. Furthermore, test data can be sent elsewhere for analysis if analysis is not presently available or stored. Thus, through connection with external control, monitoring, and analysis devices, the network analyzer has become one of the most powerful measurement tools available today.

Specifications

Network analyzers are expensive instruments. But they are of high quality and have many options. For the two applications just described, two specifications are significant: amplitude accuracy and phase accuracy. Amplitude accuracy is usually a function of range and is typically less than ±1%. Phase accuracy is typically less than ±1 degree. An internal calibration signal is usually available for checking the gain and the phase of both input channels.

Control inputs are usually internal keyboard and IEEE–488 or RS–232 ports. Data output is generally on an LED or LCD display and, occasionally, a data printer. A graphic printer or CRT can also be connected to an output port.

Measuring capabilities vary with options but generally include phase difference between inputs, phase difference between harmonics at one input, and amplitude difference between two inputs. Other modes are power, volt-amperes, power factor, harmonic distortion, impedance, frequency, voltage ratio, and transfer function.

SUMMARY

A spectrum analyzer is a frequency-selective voltmeter. It is used to measure the characteristics of components, circuits, and systems as a function of frequency. Typical applications are the measurement of the frequency response of an amplifier and the quality of a filter. Another example is the evaluation of oscillator output. In this case, a voltmeter could measure the output amplitude and a counter could measure the frequency. However, neither device would be likely to indicate the presence or magnitude of harmonics. A spectrum analyzer could do just that.

There are three basic spectrum analyzer systems: sweep frequency, multiple-filter, and time compression. In the sweep frequency system, a signal to be evaluated enters an amplifier-attenuator circuit and is changed to a workable magnitude. From there, it passes on to a mixer, where it mixes with a signal from a voltage-controlled oscillator. The output of this stage is sent to a low-pass filter, where the sum is removed. The difference goes on to a detector. The detected signal is filtered once more and passed on to the vertical input of a CRT.

The horizontal sweep of the CRT is provided by the ramp from a sweep generator, the same generator that controls the voltage-controlled oscillator. This system produces a CRT display that is proportional to the amplitude of the input signal as a function of frequency. Whereas the typical oscilloscope display is in the time domain—that is, voltage versus time—this display is in the frequency domain, or voltage versus frequency.

The multiple-filter system uses parallel-connected, narrow-band filters in the input circuit. Each filter is tuned to a segment of the operating band. The signal to be evaluated enters all filters, and a high-speed scanner selects one output after another and sends it to a filter, a detector, a filter, and finally, a display.

In the time compression system, the incoming signal is scanned, segmented, converted from analog to digital data, and stored in memory. After a predetermined number of samples have been digitized and the data have been summarized, memory is read. The data are then converted back to analog and are passed on to the mixer of a sweep frequency system. The time compression system presents a composite display based on many, fast time scans.

Spectrum analyzer specifications include frequency range, frequency accuracy, resolution, amplitude range, and amplitude accuracy. The frequency range can be from 0.01 hertz to 1200 gigahertz (but not on one instrument). These instruments are generally AF, RF, or microwave frequency devices. Frequency accuracy is described as the accuracy between two points on the display and is typically from 2% to 10%. Amplitude range generally extends over 50 to 70 decibel-milliwatts and can be as low as −117 decibel-milliwatts and as high as +30 decibel-milliwatts. Amplitude accuracy is typically ±1 decibel.

Spectrum analyzers are used to measure the characteristics of both signals and systems as a function of frequency. In signal analysis, the signal's composition is determined—that is, its fundamental frequency and the frequency and magnitude of its harmonics are measured. When a system or device is analyzed, its frequency response is determined. A constant-magnitude, varying-frequency signal is applied, and the output as a function of frequency is determined.

A network analyzer has all the capabilities of a spectrum analyzer. It can also excite the input of a system, monitor its output, and calculate amplitude and phase differences. One common measurement is determining the impedance of a component. In this situation, a constant voltage is applied across the component and swept across a frequency band. This voltage is also applied to the input of the analyzer as a reference. With the aid of a current transformer in series with the component, a voltage proportional to current is obtained and applied to the other input of the analyzer. These two signals are then converted from analog to digital data and stored. Upon completion of a programmed number of sweeps, the data are removed from memory and processed. The quotient of the two values is the impedance of the component.

The typical network analyzer has IEEE–488 or RS–232 ports, which allow attachment to a data bus. Thus, the analyzer can be externally controlled with complex test routines, the sequence can be monitored, and data can be analyzed more extensively and stored with the aid of remote computers.

BIBLIOGRAPHY

Spectrum analyzers adapt to diverse tasks, Phil Feinberg, Electronic Design, October 11, 1980.

RF and microwave spectrum analyzers, Warren Yates, Electronic Products, June 8, 1982.

Analyzers, Charles H. Small, EDN, February 9, 1984.

QUESTIONS

1. Sketch the block diagram of a spectrum analyzer, and explain the function of each section.

2. Name four specifications of a spectrum analyzer, and describe typical values.

3. Compare the operation of the three basic spectrum analyzer circuits.

4. Explain the advantages of a spectrum analyzer over an oscilloscope for the measurement of voltage.

5. Explain the advantages of a spectrum analyzer over a counter for the measurement of frequency.

6. Describe the purpose of a tracking generator.

7. Discuss the limitations of a sweep frequency spectrum analyzer.

8. Sketch the block diagram of a network analyzer, and describe the function of each section.

9. Name two specifications of a network analyzer, and describe typical values of each.

10. Compare the operation of a spectrum analyzer and a network analyzer.

11. Compare the capabilities of a spectrum analyzer and a network analyzer.

12. Describe how a spectrum analyzer can be used to evaluate a transformer.

13. Describe how a network analyzer can be used to measure frequency response.

14. Describe how a network analyzer can be used to measure the impedance of a component.

15. Discuss the advantages of time compression over real-time analysis.

16. Discuss any possible problems a multiple-filter system might have.

17. Explain the difference between frequency domain and time domain.

18. Describe how you could evaluate the performance of a communications receiver with a network analyzer.

14

Automatic
Test Equipment

OBJECTIVES

This chapter introduces some of the equipment, capabilities, and procedures in automatic testing and measurement. While this area of electronics technology is undergoing rapid growth and change, its basic role and concepts can be described.

Compare manual and automatic testing.
Describe the capabilities and control of two types of robots used in automatic testing.
Describe the instrumentation used in automatic component testing.
Describe the characteristics and applications of the IEEE–488 bus.

BASIC CONCEPTS OF AUTOMATIC TESTING

Automatic test equipment (ATE) is rapidly becoming dominant in high-volume initial and final test situations. Thus, the future technician may wonder, "Will there still be a job for me?" There will continue to be jobs for electronics technicians, for someone is needed to build, test, install, program, and service the ATE instrumentation. Therefore, this section introduces some basic concepts of ATE.

Manual versus Automatic Testing

In a typical manual test situation, a technician is provided with an item to perform measurements on, the appropriate instruments, specifications, and a given amount of time in which to accomplish the test. The technician's task is to conduct appropriate measurements, compare the results with the specifications, and make a decision. Is the item within specification? Is it good or bad? Should more items like this one be manufactured?

In a typical automatic test situation, all of the preceding tasks can be done automatically. The item to be tested is connected to instruments, which, in turn, are connected to a microprocessor-based controller. The controller commands specific signals to be applied and responses to be measured. These responses are then compared with internal standards, and a good-bad or pass-fail decision is made. Most ATE systems also provide summary data about hourly or daily batches, including the number of good or bad items and the mean and the standard deviation of each parameter measured.

Initially, ATE systems were only used in high-volume production applications, because of their high cost. Now, their use is becoming more widespread since they are cost-effective and flexible. Cost-effectiveness is a result of many factors. Some of these factors are as follows: ATE is much faster than a human, does not get a headache, needs no break, can work 24 hours per day, and asks for no fringe benefits. Furthermore, ATE does not get distracted and make errors. Finally, ATE input devices can be placed in environments unacceptable to humans.

Range of Testing Capabilities

All ATE systems can evaluate discrete and integrated components, circuits such as those on printed circuit (PC) boards, and complete systems. They control inputs, monitor outputs, compare results with standards, make pass-fail decisions, and analyze and summarize the results. In addition, ATE systems can interact with other systems that move the items to be tested in and out of fixtures. In other words, the ATE can connect, measure, analyze, decide, and summarize all by itself.

Typical Automatic Test System

Figure 14–1 shows a typical automatic test system for PC boards. A finished board arrives at the test area on a conveyor. A *robot* grips and then transports it to the test fixture. Signal sources commanded

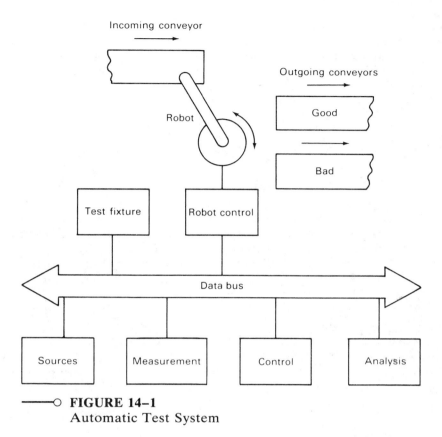

─○ **FIGURE 14–1**
Automatic Test System

by the controller excite the board. Responses are measured by instruments and sent on for comparison and analysis. The tested board is then delivered by the robot to the good or the bad conveyor and sent on to the next appropriate station.

Usually, the robot uses a two-sided grip and has the following routine. It picks up an incoming board (board 2), goes to the test fixture, picks up the tested board (board 1) with the second grip, rotates the grip and inserts board 2 into the fixture, delivers board 1 to the outgoing conveyor, and then gets board 3 from the incoming conveyor. Board 2 is tested during the 10 or 20 seconds this sequence takes.

The number of tests that can be performed in this period of time depends on the types of tests conducted, the test object, and the instrumentation. Hundreds or thousands of tests are not unreasonable. Compare that amount with the one test a technician might be able to do in the same period of time. Of course, all systems do not use robots for pickup and delivery. Many bench-top ATE systems use people to insert and remove the objects. And automatic-insertion machines are often used for discrete components. However, robots are becoming a major part of the system, and, therefore, they are included in the discussion here. The other parts of an ATE system, instrumentation and interfaces, will be described shortly.

HANDLING OF TEST OBJECTS

Parts or boards to be connected with ATE systems can be placed by hand, by automatic feeders, or by robots. Figure 14–2 shows a chip handler, which is a microprocessor-controlled system for automatically inserting leadless chip carriers (LCC) in test sockets and then removing them when the tests are complete. Used as part of a test system, the chip handler delivers and postsorts components at the rate of one per second, exclusive of testing time. Testing and analysis are conducted by other parts of the system.

Robots are common in circuit and board handling since they can also be programmed to effectively execute an otherwise boring and repetitive routine. An advantage of robots is the extensive and variable range of motion available. Typically, a robot has its own program, controller, and interface bus, and it only needs a command from the ATE system to begin operating.

Types of Robots

One definition of a robot describes it as a reprogrammable, multifunctional manipulator. In ATE applications, robots are generally dedicated; that is, they have one location and one assignment. Typically, that assignment is to move objects to be tested from one location to another through a programmed routine. They are categorized by "how" rather than by "what" they do.

Robot categories include power source, method of motion, number of motions, and method of control. Large industrial robots use pneumatic or hydraulic power to produce motion. Therefore, a compressor or hydraulic power supply is required. Smaller robots often use readily available electricity. Trends today are toward hydraulic- and electric-powered robots rather than pneumatic-powered robots.

In addition to having a power source, a robot has a *manipulator* (arm) with a gripper on the end. There are two basic forms: cylin-

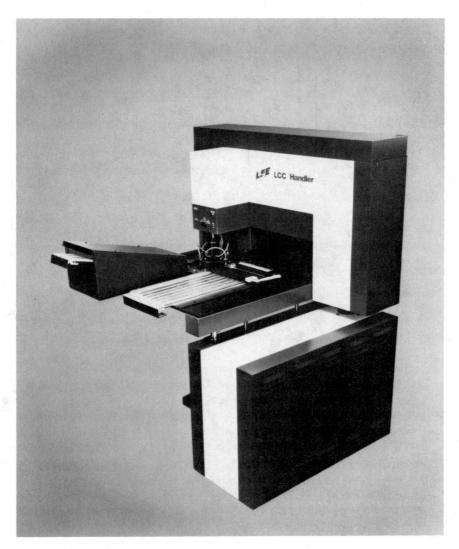

───○ **FIGURE 14–2**
Chip Handler (Photo courtesy of LFE Corporation)

drical and jointed. The cylindrical manipulator in Figure 14–3A is typically pneumatic or hydraulic, is fast moving, and consists of concentric cylinders. The jointed manipulator in Figure 14–3B is often electrical, is more complex, and provides a wider range of manipulator motion than the cylindrical manipulator. While ATE applica-

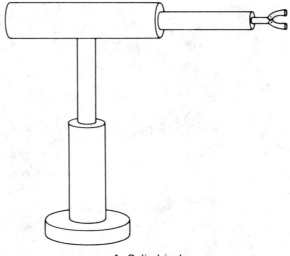

A. Cylindrical

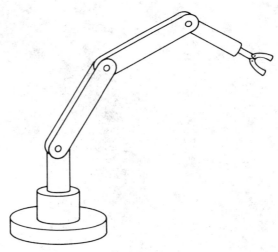

B. Jointed

─────○ **FIGURE 14–3**
Manipulators

tions typically have a gripper at the end, many different appendages are possible. In automobile assembly, for example, the appendage could be a spray gun or a spot welder.

Manipulator Motion

The way a manipulator moves can be described as point-to-point or continuous motion. Moving a component from a conveyor to a test fixture involves point-to-point motion, while spray painting the side of an automobile involves continuous motion. Point-to-point motion is accomplished by a cylindrical manipulator, and continuous motion requires a complex, jointed manipulator.

Figure 14–4 shows the simple motion of a cylindrical manipulator. An automatic test, fetch-and-insert activity, for example, would require the manipulator to be rotated to the proper direction, its gripper elevated and extended to the object source, and its gripper closed to hold the object. Then, the manipulator elevation, extension, and rotation would be changed to move the object to its destination. Next, the gripper would be opened, the object released, and the cycle repeated.

Each segment of the manipulator in Figure 14–4 has one degree of freedom; that is, each segment can move in only one plane. One segment can extend or contract, one can go up or down, and one can rotate. The gripper can only open or close. While the combi-

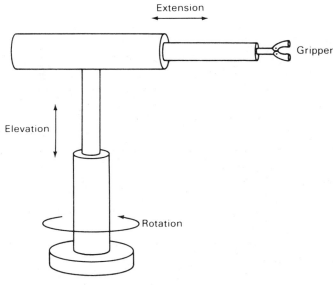

———○ **FIGURE 14–4**
Simple Manipulator Motion

nation of these capabilities does provide a wide area of gripper positioning, the resulting four degrees of freedom is somewhat inhibiting.

Consider the jointed manipulator in Figure 14–5, with the five degrees of freedom shown. One significant advantage of this manipulator is that the gripper attitude can be varied over the area of motion. It can reach up and down or from left to right. This system also has a greater reach than the manipulator in Figure 14–4, since it is not limited to one-half of the cylinder length. Adding more joints increases the flexibility of reach of this manipulator. However, with increased joints, control becomes much more complex.

Methods of Control

The control of a robot includes indicating where it is to go, indicating how it will get there, and recognizing when it has arrived. For control in a cylindrical robot, air or hydraulic fluid from a tank or a pump is directed by valves into the ends of each cylinder, causing their pistons to move out or in. Preset pins stop the travel at the desired limits. System pressure is sufficient to produce cylinder motion, but it is not enough to override the preset stops. Also needed are commands for the valves, which will be described shortly.

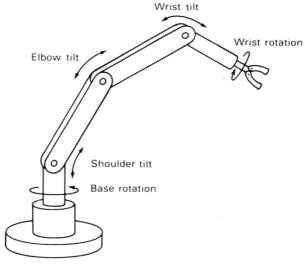

─────○ **FIGURE 14–5**
Complex Manipulator Motion

In the simple system just described, the control system needs to know only the direction of desired motion in each degree of freedom. The control of a more complex, jointed system needs to know where the robot is at all points along the way. For this more complex control, a servo system is used. Sensing transducers are attached to each joint through gears or other mechanical linkages. The transducers are potentiometers or LVDTs in analog systems and interrupted light beams in digital systems. The light beam passes through a hole in a disk attached to the joint's drive motor (the actuator) and produces one pulse per revolution.

Thus, the position of each joint is controlled by the magnitude of a linear voltage in analog systems and the number of pulses in digital systems. This feedback information is compared with a control signal, producing an error. As described in earlier chapters, this error signal indicates both the magnitude and the direction of the error between where the robot is and where the robot should be. This signal is then amplified and sent on to the control valves for the pistons, the windings of a DC servo motor, or whatever the system uses for an actuator.

Programming

A simple cylindrical system with pistons can be programmed with a rotating drum, similar to the drum in a music box, as shown in Figure 14–6. The drum is rotated by a motor, through gears. Pins are placed in the drum and actuate switches (such as open and close in the figure) as they pass by. Each direction of motion has a switch, which,

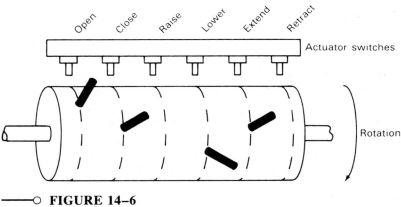

─○ **FIGURE 14–6**
Cylindrical Programmer

in turn, commands an actuator. Programming consists of placing the pins so that they command specific directions of motion in relation to others and according to elapsed time.

The servo systems used in jointed robots require more complicated programming. A servo system is likely to be digital and to be controlled by a microprocessor. Hexadecimal operation codes (op codes) command the direction of motion for each degree of freedom, and hexadecimal numbers represent the magnitude of motion. As mentioned earlier, magnitude of motion can be measured by a count of light pulses. Programs, which are read from memory, command the motions in a predetermined sequence. Errors between the desired and the actual positions are translated into actuator commands.

Programs for complicated systems can be generated in two ways. First, motions can be mathematically described and translated into op codes. Second, and more commonly, the system can be patterned. In this approach, the system is manually driven through one cycle and instructed to remember each significant position—that is, where the arm stops, when the gripper opens, and so on. This program can then be stored for future use in some type of memory device.

AUTOMATIC TEST INSTRUMENTATION

Bench-top and floor model automatic test instrumentation is available for the evaluation of discrete components, integrated circuits (ICs), PC boards, and complete systems. Among the most common instrumentation is the bench-top, linear, IC tester; a typical IC tester is shown in Figure 14–7. This section discusses the IC tester and ATE fixtures, specifications, and interfaces.

Linear IC Tester

The typical linear IC tester can perform a battery of tests on a wide range of IC types and models. These circuits include operational amplifiers, regulators, oscillators, comparators, ADCs, DACs, and voltage followers. Discrete components such as diodes and transistors can also be tested. The range of tests that can be conducted is also quite extensive. For example, the following tests can be automatically conducted by the typical ATE system on an IC op amp:

Gain,
Slew rate,
Bias current,

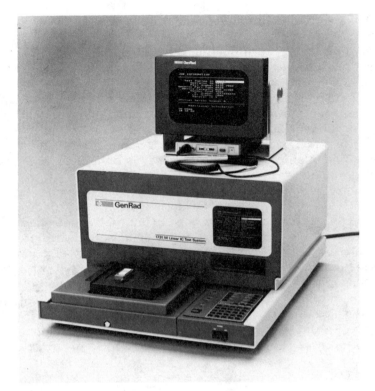

——○ **FIGURE 14–7**
Linear IC Tester (Photo courtesy of GenRad, Inc.)

Offset voltage,
Offset current,
Power supply current,
Output voltage swing,
Gain-bandwidth product,
Common-mode rejection ratio.

To use a linear IC tester, the operator programs the instrumentation. Programs are entered with a keyboard by following a
menu and prompt instructions that appear on a CRT monitor. Programming consists of specifying the type of component, the test conditions, the magnitudes of the independent variables, and the
acceptable range of results. Components can be automatically categorized in the testing process as good or bad or within some specified boundaries.

Once a program is entered, the ATE system conducts unlimited testing. Components are placed in the test fixture by hand, by using a robot, or by using an automatic handler. Removal follows the same procedure, with automatic binning by test results being common. That is, each component is directed into one of a number of chutes, holders, or bins, according to selected parameters. For example, rejects may be sent in one direction and good items in another. Or rejects may be further categorized by specific defect, and good items may be classified by ranges of a selected parameter. Most automatic test equipment allows the operator to select binning parameters and establish boundaries.

Figure 14–8 outlines the basic sections of a simple linear IC tester. The *device under test* (DUT) is held in a test fixture. Input and feedback resistances are among the internally controlled parameters, as are $+V_{CC}$ and $-V_{CC}$. Input signals come from either a sinusoidal or a complex waveform generator. The selection of the specific source, the frequency range, the sweep rate, the sweep range, and the amplitude are determined by the controlling microprocessor.

The output of the DUT is connected to AC and DC voltmeters (VMs), which, in turn, are connected to the microprocessor analysis circuitry. In the microprocessor, these outputs are compared with input references for the computation of gain, frequency response,

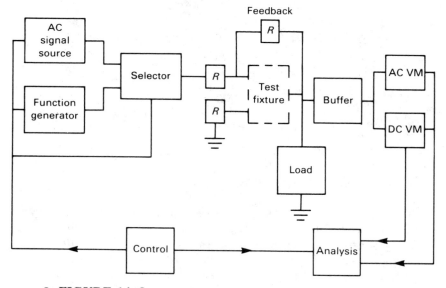

─○ **FIGURE 14–8**
Sections of Linear IC Tester

and so forth. In addition to categorizing specific components as good or bad, ATE systems provide output data. Individual and/or summary test results can be sent over an IEEE–488 data bus to a computer.

Fixtures

While some ATE instruments are designed for a specific type of component, many handle a wide variety of components. One input system provides a *bed-of-nails* input fixture with socket adapters, as shown in Figure 14–9. Fixtures are devices that align, hold, or otherwise provide positioning and interconnection between components to be tested and the testing system. The bed of nails offers a universal set of connections that can be adapted to a wide range of components. For the testing of a specific model of component, an adapter is added. These adapters serve as connector jumpers and masks. That is, they cover most connections while allowing access to specific connection points, according to the type of device. For example, an

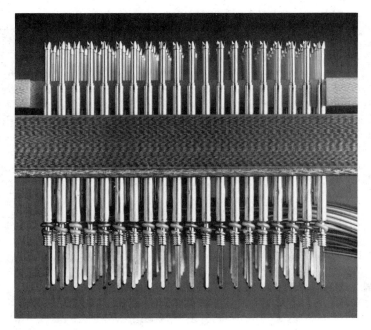

─○ FIGURE 14–9
Bed-of-Nails Fixture (Photo courtesy of GenRad, Inc.)

op amp adapter board provides access to the appropriate ATE input pins for a wide range of op amps.

Adapter boards are available for other families of devices, such as regulators, converters, and discrete components. They speed up the testing process and reduce operator error or misplacement. When adapter boards are combined with an automatic feed system, the testing process is even faster, and the operator is eliminated.

Specifications

Generally, ATE systems are described by type of components tested, type of control, data analysis capabilities, communication interfaces, and control and accuracy in input and output parameters. Typical categories of component testers are analog IC and discrete components, digital IC, in-circuit boards, and memory testers. Control is usually accomplished with a microprocessor, using a program that is stored on board in a read-only memory (ROM), entered through the keyboard, or sent over a control bus.

Test data analysis can vary from categorizing specific devices as good or bad to developing curves that show the number of units as a function of magnitude for many of the test parameters. Most often, though, test data are stored internally and sent over a data bus to be processed elsewhere. The most common data bus is the IEEE–488, which is described shortly.

The range and accuracy of input and output magnitudes are too involved to discuss here. And there are dozens of parameters to consider for each type of ATE system. Suffice it to say that equipment is becoming more accurate as each day passes. Meanwhile, the greatest cause of error, the operator, is becoming less involved. Therefore, it is fair to say that automatic test systems tend to be more accurate than manually operated systems because of internal regulation, digital technology, and the removal of human error.

Interfaces

Most ATE systems, like many test instruments, have an IEEE–488 data bus connector. This system, also known as the *general-purpose interface bus* (GPIB), has been a standard in the test industry since 1975. It is used to communicate commands and data between instruments, sources, and other sections of an automatic test system. Many devices—both devices that send and devices that receive—can be connected to the bus at one time. Some devices, like the control microprocessor, both send and receive. The bus itself is bidirectional.

Figure 14–10 shows the connections for an interface bus. Devices are connected to the bus through a 24-pin connector. Of these connectors, one is a braided shield and is used as the system ground, 7 are signal grounds, and the remaining 16 are used for information. Twelve of the 16 information connectors are in twisted pairs. The *American Standard Code for Information Interchange* (ASCII) is commonly used in these systems.

SUMMARY

Automatic test equipment (ATE) is becoming increasingly more common in high-volume, initial and final test applications. It is used to evaluate and categorize components, circuits, and systems as good or bad or within a specific band of performance criteria. While ATE usually has a high purchase price, it pays for itself by working long hours without a break, working quickly, and making few errors.

The typical ATE system usually moves the device under test (DUT) in and out of a test fixture, controls the operating power and the signals for the DUT, evaluates the output, determines conditions, and summarizes individual and batch DUT results. Handling can be done by hand, by robot, or by other automatic-handling sys-

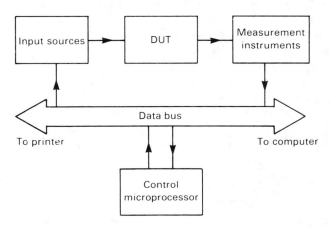

FIGURE 14–10
Interface Bus

tems. Signals come from sources such as sinusoidal or function generators, while power comes from regulated supplies. Control and analysis are accomplished by one or more microprocessors, and results are communicated externally by an interface bus.

Robots for handling parts can be simple cylindrical types or more complex models with jointed arms (manipulators). Manipulator power can be electric, hydraulic, or pneumatic. Methods of control vary. At one end of the spectrum, cylinders are extended and retracted against travel-limiting stops. Power to each cylinder comes from the source through valves that are operated by a pin-drum system similar to the drum in a music box. At the other end of the spectrum, servo-controlled robots have digital or analog transducers at each point of movement. The output of each transducer is compared with a command signal from a stored program. Robot motion is a result of error between the actual and the desired position for each degree of freedom and is produced by motors, pistons, or other actuators. Degrees of freedom describe the number of planes involved in manipulator movement. For example, for a piston, up/down movement is one degree of freedom, and extend/retract movement is another degree of freedom.

The instrumentation in ATE systems (floor or table models) is usually self-contained. For example, a linear IC tester includes power and signal sources, adjustable source, load, and feedback resistances, output-measuring instruments, data and program memory, and a microprocessor. This system could measure gain-bandwidth product, slew rate, output voltage swing, common-mode rejection ratio, and many other specified parameters. Other ATE systems are available for digital ICs, PC boards, and complete circuits.

A DUT is usually connected through a bed-of-nails, multipin fixture over which a mask is placed. The mask has an opening over the appropriate pins and has jumpers underneath for the needed internal connections. With this fixture, manufacturers can produce a system that can test a wide range of linear ICs by family, such as op amp, regulator, converter, diode, or transistor.

Specifications for an ATE generally describe type of components tested, type of control, data analysis capabilities, communication interfaces, and control and accuracy of input and output parameters. Accuracy is typically better than accuracy for manually operated systems because of internal regulation, digital technology, and the removal of human error. Interfacing is almost always accomplished with an IEEE–488 bus. This bus uses 24 conductors to both send and receive commands and data between sources, instruments, and other peripheral systems.

BIBLIOGRAPHY

Benchtop ATE system operates at near production speed, EDN, July 21, 1983.

Analog/linear IC testing—flexible test systems solve complex challenges, Donald R. Glancy, Test & Measurement World, September, 1983.

Board testers rise to VLSI challenge, Electronic Engineering, December, 1983.

Computer Controlled Testing & Instrumentation, Martin Collins, Halsted Press, 1983.

Introduction to In-Circuit Testing, 2nd ed., GenRad, Inc., 1984.

QUESTIONS

1. Describe three advantages of automatic testing over manual testing.

2. Name three categories of devices commonly tested with ATE.

3. Name three sections of an ATE system.

4. Describe the capabilities and control of a cylindrical robot.

5. What limitations does a cylindrical robot have?

6. Describe the capabilities and control of a jointed-arm robot.

7. What is meant by *degrees of freedom?*

8. Compare servo and nonservo control systems for robots.

9. Describe digital methods of position identification.

10. Describe how a point-to-point robot is programmed.

11. Describe how a continuous-motion robot is programmed.

12. Describe how a robot can be programmed by patterning.

13. Name five devices that can be tested by a typical linear IC tester.

14. Name six parameters of an op amp that can be tested by a typical linear IC tester.

15. Explain the advantages of a bed-of-nails fixture.

16. Compare the capabilities of an interface bus such as the IEEE–488 with the capabilities of simple input or output cables.

17. Describe two forms of binning and how they can be used.

18. Explain how a manufacturer can utilize ATE summary data after testing a large group of op amps.

15

Instrument
Calibration

OBJECTIVES

This chapter introduces calibration of electronic instruments. Some requirements, concerns, and procedures for calibration are discussed.

> Define the term *calibration*.
> Describe some methods used for electronic instrument calibration.
> Name and describe some methods of documentation.
> Describe calibration services available from the NBS.
> Describe the role of the technician in calibration.

THE CALIBRATION PROCESS

Calibration Defined

Calibration is both a condition and a process. In the first case, calibration is the condition of an electronic instrument when its output or response is as close as possible to the specified value. For instance, an oscillator is calibrated when its output frequency is as close to specification as it can be. A voltmeter is calibrated when it indicates the true value as closely as possible.

In the second case of the definition, calibration is a process of examination. To calibrate an instrument means to determine how close to the true value its indication or output really is. This process also includes the correction and/or documentation of degree of error. More will be said about documentation later in this chapter.

Parameters

Calibration is part of all international measurement. It includes the areas of optics, thermodynamics, mass, hydraulics, acoustics, aerodynamics, and many other domains. As mentioned in the early chapters of this book, all things that can be measured have standards. Likewise, all things that have standards can be compared with those standards. Comparison with standards is the process of calibration.

In the electronics industry, both sources and measurement instruments are calibrated with standards. The most common parameters that are involved are voltage, current, resistance, inductance, capacitance, frequency, and time. Other parameters include impedance, power, and electromagnetic radiation. In a nutshell, if something produces or measures a signal, it can be calibrated.

Degree of Accuracy

Accuracy is a cost-benefit issue. The establishment and maintenance of highly accurate equipment is very expensive. For this reason, companies obtain and use equipment whose accuracy is no greater than required. In other words, you do not purchase a high degree of quality if you do not need it, and you only use the best instrument when it is absolutely required. Regardless of the quality, however, you should ensure that the instrument is doing its best.

Properly working electronic instruments are capable of operating at a higher quality than their specified levels. For example, a 2% FSD voltmeter will usually give readings much closer to the true value than the 2% accuracy indicates. The 2% specification represents the worst case one can expect. However, the instrument is considered to be in calibration if its indication is within the specified 2% range. It is out of calibration if it is not within the specified range. As another example, a 0.01%, 1-megahertz oscillator is within calibration if its output frequency is between 999,900 and 1,000,100 hertz.

In summary, an instrument should be at least as accurate as its specified value. If it is not, some form of compensation is required, such as correction. But regardless of what compensation steps are taken, a calibrated instrument always requires documentation.

Documentation

Paperwork and documentation are integral parts of the calibration process. Documentation comes in many forms. The most common form is a tag or label, such as the one shown in Figure 15–1. This tag, which is attached to the instrument, generally indicates the instrument's identification number, the date of calibration, who performed the calibration, when recalibration should occur, and a good/ bad comment. Many organizations also maintain a calibration book, which indicates, by serial number, when each instrument was calibrated and by whom, the adjustments made, and other appropriate information.

Other forms of calibration record are a graph, a table, or a statement of instrument performance. For example, a graph could present error as a function of indicated value for a voltmeter scale. Data could be from ten or more equidistant points of comparison, over the full-scale range, between the voltmeter and an identified standard. Figure 15–2 shows a sample graph for voltmeter error. In

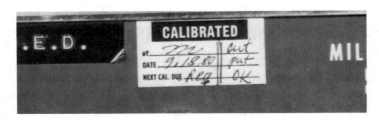

———○ **FIGURE 15–1**
Instrument Calibration Tag

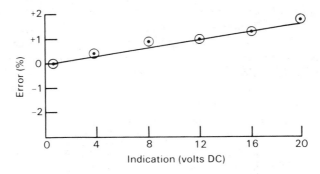

———○ **FIGURE 15–2**
Voltmeter Error

this example, the graph shows that the meter indicates a higher value than the true value at an amount that increases with the input voltage. Thus, when this meter indicates 12 volts, there is a $+1\%$ error. So, the technician must reduce the indicated value by 1%, or 0.12 volt, to determine the true value.

The graphing approach is taken when, for some reason, correction is not possible. As mentioned in an early chapter, if you know the magnitude of an uncorrectable error, you can compensate for it in your data. So, records let the user know an instrument's most recent date of calibration, whether it is acceptable, its degree of variation, and when its calibration is due next.

A calibration process that once was considered only helpful is now quite often a requirement for military specifications. This process is traceability. *Traceability* is defined as the ability to trace an instrument's calibration back to a reference to the National Bureau of Standards (NBS) through an unbroken chain of documented calibrations. In manufactured products, traceability means that the values of all specific components must have qualifying paperwork. The instruments with which they were measured must have documentation. Performance of the finished product must have been measured by instrumentation having documented performance. In summary, all quantification of components or product performance by a supplier must be supportable to the highest level, which is the NBS. While traceability is often required, it is a costly and time-consuming process.

Frequency of Calibration

How often should an instrument be calibrated? There is no absolute answer. But a simple answer is, Calibrate an instrument often enough to ensure that it is within specifications. Calibration can be expensive because of the downtime for the instrument and the cost of the calibration service. For this reason, companies may not wish to calibrate instruments very often. Thus, they may calibrate them only at an interval required by their customers, such as the military. Or they may calibrate them according to a predetermined company policy, only after repair, or when they think that the instrument is not performing well.

Most companies have some kind of calibration schedule, and it is generally based on experience. For example, if you check a voltmeter every week for two years, you will discover that it does not change. It will probably remain within specification for many months, unless someone drops it or subjects it to some other abuse. Checking this instrument once a month or every three months would probably be a reasonable policy. However, an instrument with a history of slow or sudden change should be checked more frequently. For instance, battery-operated instruments need frequent checking.

While there is no absolute industry standard for time intervals between calibration, there are some averages. Typical calibration intervals for counters, multimeters, and oscilloscopes are 3, 6, and 12 months. Most companies that are free to select their own interval calibrate their instruments every 6 months.

CALIBRATION SERVICES

Professional calibration services can be obtained from the National Bureau of Standards, a company specializing in calibration services, or an in-house technician. Available services include performing calibration, teaching others how to calibrate, and providing calibration signals. Our discussion begins at the highest level, the NBS.

National Bureau of Standards

The NBS offers calibration services to the military, industry, schools, and any individual who is interested in quality measurement. Services include calibration of primary standards, standard reference data, measurement seminars, technical information, and publications.

A company obtains calibration generally by sending its precision measuring instrument or standard to the appropriate NBS facility. Of course, a device sent to the NBS should have a quality that warrants this level of evaluation. Then, arrangements must be made with NBS, and the instrument must be properly packed and safely delivered. Since the NBS does not repair instruments, the instrument must be in proper working order, with connections clean, fresh batteries installed, and so forth. The calibration process at the NBS may exceed a month, depending on its current backlog.

The actual calibration activity depends on the item sent. For example, a precision oscillator may be calibrated at frequencies of 0.1, 1.0, 5.0, and 10 megahertz, with a reference standard accuracy of the order of one part in 10^{13}. A report is then sent along with the device when it is returned. This report contains the measured values and their degrees of uncertainty. The report, however, cannot describe later long-term drift, effects of return transportation, or effects of operation under conditions unlike those of the calibration situation. Also, the NBS does not recommend a specific recalibration schedule.

Other calibration-related services of NBS include measurement seminars and workshops. These services are available, at a cost, to a limited number of persons working in measurement or standards labs, who meet prerequisites relating to education, work experience, and current job. The seminars last from one to five days and include lectures, demonstrations, and discussions.

Consultations on measurement problems are available over the telephone or through the mail from the NBS. The NBS also offers a wide range of publications resulting from research and describing services. Finally, the Standards Information Service maintains a reference collection of more than 260,000 engineering standards issued by U.S. technical societies, professional organizations, trade associations, government agencies, international standardizing bodies, and so on.

Industrial Calibration Services

Most industrial communities have private companies that specialize in the repair and calibration of electronic instruments. Some specialize in one brand, but most companies service a wide range of brands and types of instruments. Whether you take the instrument to them or they come to you depends on the type of instrument and the quantity to be calibrated. Typically, these organizations inspect, calibrate, and align (if appropriate) the instrument and return it with a calibration report. Like the NBS reports, their report describes error between your instrument and their standards. They will also compare their standards with NBS standards for traceability, if required.

Calibration Technicians

Many electronics firms have their own calibration department. Generally, this laboratory is a clean room that is dust-free, with the temperature and humidity closely controlled. Quite often, the area is electromagnetically shielded. In this room, instruments used in production, test, and quality control are calibrated according to a predetermined schedule.

The persons performing the calibration task are generally calibration technicians. See Figure 15–3. Also called calibration lab technicians or instrumentation technicians, they generally have a technical school or community college education in this specific area of electronics. Typically, they also have patience and a great respect

──○ **FIGURE 15–3**
Electronic Technician Calibrating Instruments (Photo courtesy of Ametek, Inc.)

for quality. Their primary responsibility is to inspect, calibrate, and repair electric, electronic, and mechanical measuring and indicating instruments for conformance to established standards. Calibration technicians also write reports of calibration activities, maintain calibration records, suggest better ways to ensure instrument quality, and maintain the company's standards lab.

CALIBRATION CHECKS

All instruments should be inspected and calibrated by professionals on a regularly scheduled basis. There will be times, however, when the instrument you are using will have missed its scheduled recalibration, and recalibration is not possible at the moment. Yet, you question the accuracy of the instrument. In this situation, there are some calibration checks that you can do.

Technician's Limits

It is reasonable for you as a technician to perform some test to check the accuracy of an instrument. It is also reasonable to document the results and consider them in your measurements. However, corrective steps should not be taken. Since there is always the possibility that you might make matters worse, correction should be left to a calibration technician.

Thus, as a technician, you can check, document, compensate, and recommend, but you cannot correct. For example, suppose you measure the output of a variable-voltage DC power supply with an accurate meter and determine that there is little relationship between the dial setting and the output voltage. The voltage is well filtered, stable, and adjustable, but the indicated value is wrong. You can compensate for this error by keeping a voltmeter attached or by developig a curve that shows the error between indicated and actual values.

Some sample error graphs are presented in Figure 15–4. They show that the error may not be constant over the full range of a scale. Figure 15–4A shows an error that is relatively constant, about 1% high over the full range of the input. Constant error can often be corrected with readjustment of the indication at full scale. The linear error in Figure 15–4B increases progressively as the input increases, beginning at 0% at 0 volts input and gradually rising to above 2% at full-scale. Although error like this can also be corrected with internal readjustment, the graph describes the error and allows for compensation until recalibration is achieved. The error shown in Figure 15–4C is nonlinear, varying in both magnitude and direction. Usually, an error graph is the best way for a technician to deal with nonlinear error.

The process just described is a common way to compensate for error in sources or in measuring instruments. It is noninvasive since the equipment is not opened, modified, or otherwise tampered with. And it provides you with a temporary accommodation of error while you wait for the instrument to be corrected. That is, if you know the amount of error that exists, you can indicate the error in your data and compensate for it.

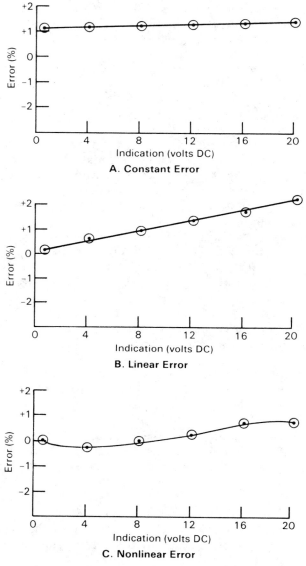

A. Constant Error

B. Linear Error

C. Nonlinear Error

FIGURE 15–4
Sample Error Graphs

Multimeters

Multimeter calibrators are available that quickly and accurately calibrate both analog and digital multimeters, the latter up to 7 1/2 digits. They are accurate, precise, self-calibrating, and quite expensive. Figure 15–5 shows a calibrator used for calibrating multimeters. It provides continuously variable but accurately known DC voltages from less than 100 microvolts to more than 1 kilovolt. Also available are DC levels from less than 100 microamperes to more than 1 ampere. Resistances available range from less than 1 ohm to more than 10 megohms.

With a calibrator like the one of Figure 15–5, a technician need not assemble a configuration of sources, loads, and various standards. Instead, the calibrator provides all these functions. But if you do not have such a calibrator, you can still perform some tests on your multimeter, as illustrated in the following examples.

○ Examples of Checking Voltage

The best way to check a voltmeter is to apply a known voltage across it and read the indication. A battery cannot be used as a DC voltage standard because of the unpredictability of its voltage. Instead, a precision power supply with a built-in voltage indication is required for DC, and a high-quality oscillator is needed for AC. The mechanical zero should be properly adjusted first. Then, six to ten points on the scale should be compared with the source.

Zero and full-scale voltages are not sufficient checks, because the deflection could be nonlinear, giving undetected midscale errors. This situation was shown in Figure 15–4C. Data should be recorded, dated, and signed by you, and error beyond the specified accuracy should be reported for corrective action. This process is repeated for all voltage ranges. An error graph should then be drawn to pro-

────○ **FIGURE 15–5**
High-Speed Multifunction Calibrator (Photo courtesy of Datron Instruments, Inc.)

vide future instrument users with a clear description of an instrument's accuracy.

A precision voltage divider can be used if the power supply is not adjustable. If a precision divider is not available, a fixed resistive divider or an adjustable divider (potentiometer) can be used. In the latter case, another voltmeter of known and higher accuracy is used. The more accurate meter is used as a standard and is connected in parallel with the meter being checked across the voltage source.

An uncalibrated source can be used with a calibrated source if it is well filtered and stable. Figure 15–6 shows the connections. The standard (reference) meter is connected in parallel. Then, the supply is varied, and the indications of the standard meter and the meter being calibrated are read and recorded.

Another approach is possible when a standard meter is not available. In this situation, you can connect a group of meters in parallel and consider the mean of their indications to be close to the true value. This procedure can only be used with meters that are operating properly and are on their best range, and the application must be within their specified capabilities.

○ **Example of Checking Current**

Current ranges are usually checked by connecting the meter being calibrated in series with an ammeter of known and higher accuracy and passing appropriate current through the circuit. The connections are shown in Figure 15–7. Thermocouple transfer in-

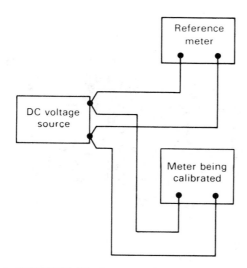

──○ **FIGURE 15–6**
Voltmeter Calibration

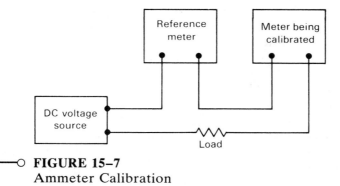

───O **FIGURE 15–7**
Ammeter Calibration

struments are commonly used in this check. A calibrated current source is ideal if it is available. Generally, a DC supply or an AC oscillator is used as a source, and a limiting resistance is placed in series.

Six to ten measurements per range are reasonable for this check. Caution should be taken with the thermocouple meter. Current should be increased gradually, because the meter responds slowly and could be overloaded before you became aware of it.

O **Example of Checking Resistance**

Resistance ranges can be checked by measuring standard resistors of the various scales. Obviously, the degree of precision with which you can do this check depends on the type of meter (analog or digital) and the section of the scale (analog). With an analog ohmmeter, zero and full-scale adjustments must be made first. Its nonlinear scale means that center is of greatest interest, and comments about the higher end are of little significance.

Almost any resistor can be used if standards are not available. You will need an accurate resistance bridge, impedance bridge, or *RLC* bridge, though. For this check, the test resistors are first measured with the bridge to determine their true values. Then, each is measured with the ohmmeter being calibrated. This check is an example of being able to trace a value back to a standard when a direct standard is not available.

Oscilloscopes

The oscilloscope is another instrument for which there are automatic calibration systems. They are computer-controlled and are programmed to check vertical and horizontal voltage calibration, rise time, synchronization, trigger, and other functions. Test results are then displayed and stored.

Since an automatic calibration system is not likely to be available to you, you will need to perform a calibration check. For instance, vertical amplifier calibration is usually achieved by placing the vertical vernier control in the calibrated position. This control may have another name, but it is the potentiometer in the center of the vertical range or volts per centimeter rotary switch. Calibration can be checked with internal signals by following instructions in the manual.

One common method for introducing internal-calibration signals is to switch the vertical range switch to the calibrated position. A square wave of specified size then appears on the CRT. In some scopes, a calibration signal is available as a front-panel test point. In either case, the square-wave test signal should produce a specific peak-to-peak deflection. If the magnitude of deflection is incorrect, adjustment is made by changing the vertical amplifier's gain potentiometer with a screwdriver.

A voltage-calibrated, square-wave generator can be used when internal signals are not available. However, you must know that the source voltage is accurate and that the square wave has low distortion. Once the vertical gain has been checked, the probe and other input impedance compensation must be checked.

Probe or other input compensation is correct when a high-quality square wave at the input is displayed with no overshoot or undershoot. Overshoot (Figure 15–8A) or undershoot (Figure 15–8B) is removed by adjusting the capacitance of the probe. This adjustment is made by rotating the probe's sleeve, as described in the probe manual or the scope manual. The correctly compensated waveform is shown in Figure 15–8C.

The horizontal sweep time can be checked by introducing a signal of known frequency and measuring its period. In many scopes, the internal vertical calibration signal is 1 kilohertz and can be used for checking the horizontal sweep calibration. Otherwise, an external signal can be used. Corrections can be made with the horizontal gain adjustment.

The oscilloscope adjustments just described are the common checks a technician can make. In contrast to the adjustments for meter recalibration, these adjustments are typically on the front panel and do not require disassembly. There are many other calibration adjustments for oscilloscopes, but they should be left for the calibration department.

Oscillators

The most important calibration check in an oscillator is for frequency, and the best way to do it is with a counter. Once again,

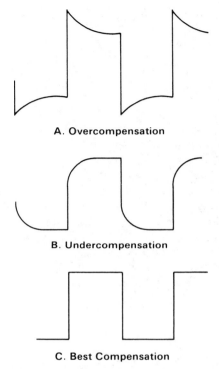

A. Overcompensation

B. Undercompensation

C. Best Compensation

——○ **FIGURE 15–8**
Effects of Probe Compensation

using six to ten points along each range is appropriate for finding absolute accuracy and linearity.

In the absence of a counter, another oscillator can be used. The oscillator being calibrated should be connected to the vertical input of an oscilloscope, and the reference oscillator should be connected to the horizontal input, as shown in Figure 15–9. The oscillator to be calibrated is then set to a frequency on the dial, and the standard oscillator is adjusted until a 1:1 Lissajous pattern is obtained on the CRT. The actual output of the test oscillator is the frequency indicated on the dial of the standard oscillator. This procedure can then be repeated at many points on this range and the other ranges. Since the oscilloscope is used only as a null indicator, it is not a factor in system accuracy.

If the standard oscillator has only one frequency, other points can be checked as multiples, again by using Lissajous patterns. For example, consider a 1-kilohertz standard oscillator and the test oscillator connected to a scope. A 1:1 pattern will appear when the test oscillator is set at 1 kilohertz. Readjusting the test oscillator

──o **FIGURE 15–9**
Oscillator Frequency Calibration

until a 2:1 pattern appears means that the ratio between the two oscillators is 2 to 1. One signal is twice the other, or one-half the other, depending on how you choose to interpret this pattern.

Another advantage of this system is that good-quality test signals are often readily available. For instance, the power line frequency in your community, while low, may be quite accurate. A frequency of 60 hertz ±1/20 is not uncommon, giving you a 0.12% standard. And as described in Chapter 2 and in the Appendix, the NBS transmits highly accurate AF and RF signals, which are available with a communications receiver.

SUMMARY

Calibration is the condition of an instrument or a source when its response or output is as close as possible to the specified value. It is also the process undertaken to ensure this condition. All parameters of international measurement are involved. Mass, thermodynamics, optics, and aerodynamics are some domains that require calibration. Some electrical parameters calibrated are voltage, current, resistance, frequency, inductance, and time.

The degree of accuracy in calibration depends on the specified quality of the source or the instrument, the level of standard available, and the price one can afford to pay. High accuracy has a high cost. An instrument can be expected to perform up to but not better than its specified capabilities.

Calibration includes documentation. A tag attached to an instrument should indicate when it was last calibrated, who performed the calibration, whether the instrument was good or bad, and when recalibration should be done. Other common records are calibration books or charts for all in-house instruments. Quite often, the calibration of components, products, or instruments must have documentation of traceability to the National Bureau of Standards (NBS).

The frequency with which instruments are calibrated may be determined by company policy, by customer specifications, or by the suspicion of problems. Many companies have a calibration schedule developed from experience with specific instruments. Calibration intervals of 3, 6, or 12 months are common.

The NBS offers a wide range of calibration services. These services include the calibration of primary standards, measurement seminars, standards reference data, consultations, and publications. Many private companies also calibrate instruments. Their standards are traceable to the NBS.

Many larger companies have their own calibration laboratories staffed by trained calibration technicians. Their duties include inspection, calibration, repair, and record keeping. Patience and an appreciation for quality are characteristics of these technicians.

There are some calibration checks that the average technician can make if calibration technicians are not available. For instance, an instrument or source is compared with a higher-quality device, errors are quantified and noted in the measurement data, and a request for recalibration is passed on to the appropriate individual. Readjustment or correction of an instrument is not recommended.

A voltmeter can be checked by measuring the output of an adjustable voltage standard at six to ten points on each range. In the absence of an adjustable standard, a fixed standard and potentiometer can be used. If a standard is not available, a smooth and stable source can be used. Its output is monitored with a voltmeter of known and higher quality than the meter being checked. Another approach is to connect a group of voltmeters in parallel and use the mean indication as the standard.

An ammeter can be checked by connecting it in series with a calibrated current source and appropriate resistance. In the absence of a calibrated current source, a good-quality source, a standard ammeter, and an appropriate resistance can be connected in series with the meter being tested.

An ohmmeter can be checked by measuring the values of several standard resistors. When these standards are not available, almost any resistor can be used if its value is first checked with a resistance, impedance, or *RLC* bridge. Of course, the meter should have good batteries and be properly zeroed.

Common calibration checks can be made of the vertical amplifier, horizontal amplifier, sweep time, and probe compensation of an oscilloscope. A calibration square wave is used, and, generally, it is internally produced by the oscilloscope. The amplifiers are adjusted to ensure proper gain. The probe is adjusted to balance the input impedance and eliminate overshoot or undershoot.

An oscillator is best checked by measuring its output frequency with a counter. Once again, six to ten points along each range are reasonable for ensuring accuracy and linearity. When a counter is not available, another oscillator and an oscilloscope can be used. Connecting the test oscillator to the vertical input of the scope and the standard oscillator to the horizontal gives a Lissajous pattern on the CRT. Then, the standard can be adjusted until a waveform with a 1:1 or other ratio is produced.

Determining the error in any source or instrument is the reason for calibration in measurement. An error that is stable and known can be accommodated in data, even if it cannot be removed. The essence of measurement is to quantify the magnitude of a parameter and the degree of uncertainty in that quantification.

BIBLIOGRAPHY

The transfer of accuracy: microwave standards from the NBS to the user, Mitchell Rose, Test & Measurement World, February–March, 1982.

Instrument calibration: how much? how often? Test & Measurement World, May, 1982.

Calibration study results favor systematic, uniform approach, Test & Measurement World, June, 1982.

How accurate is accurate enough? Warren Yates, Electronic Products, June 30, 1982.

The design of an AC calibrator, B.W. Kerridge and P.R. Harold, Electronic Engineering, January, 1984.

Calibration and Related Measurement Services of the National Bureau of Standards, National Bureau of Standards, 1980.

QUESTIONS

1. Define *calibration* as a state and as a process.
2. Define *traceability* and explain how it is achieved.
3. Describe how the frequency of calibration is determined.
4. Describe a technician's role and limits in instrument calibration.
5. Describe the qualifications and duties of an instrument or calibration technician.
6. Name and describe four services available from NBS.

7. Explain how to compensate for error when instrument adjustment or correction is not possible.

8. Describe three methods of documentation and the purpose of each.

9. Describe two methods of calibrating a voltmeter.

10. Describe one method of calibrating an ammeter.

11. Describe one method of calibrating an ohmmeter.

12. What is the best way to calibrate an oscillator? Explain.

13. Explain how to make NBS frequency standards available in your lab.

14. Explain how Lissajous patterns and local line voltage can be used to calibrate an audio frequency generator.

15. Describe how to calibrate the vertical and horizontal amplifiers of an oscilloscope.

16. Describe the purpose and the method of oscilloscope probe compensation.

Appendixes

Appendix A: Common Logarithms

Four Place Common Logarithms

N.	0	1	2	3	4	5	6	7	8	9
10	0000	0043	0086	0128	0170	0212	0253	0294	0334	0374
11	0414	0453	0492	0531	0569	0607	0645	0682	0719	0755
12	0792	0828	0864	0899	0934	0969	1004	1038	1072	1106
13	1139	1173	1206	1239	1271	1303	1335	1367	1399	1430
14	1461	1492	1523	1553	1584	1614	1644	1673	1703	1732
15	1761	1790	1818	1847	1875	1903	1931	1959	1987	2014
16	2041	2068	2095	2122	2148	2175	2201	2227	2253	2279
17	2304	2330	2355	2380	2405	2430	2455	2480	2504	2529
18	2553	2577	2601	2625	2648	2672	2695	2718	2742	2765
19	2788	2810	2833	2856	2878	2900	2923	2945	2967	2989
20	3010	3032	3054	3075	3096	3118	3139	3160	3181	3201
21	3222	3243	3263	3284	3304	3324	3345	3365	3385	3404
22	3424	3444	3464	3483	3502	3522	3541	3460	3579	3598
23	3617	3636	3655	3674	3692	3711	3729	3747	3766	3784
24	3802	3820	3838	3856	3874	3892	3909	3927	3945	3984
25	3979	3997	4010	4031	4048	4065	4082	4099	4116	4133
26	4150	4166	4183	4200	4216	4232	4249	4265	4281	4298
27	4314	4330	4346	4362	4378	4393	4409	4425	4440	4456
28	4472	4487	4502	4518	4533	4548	4564	4579	4594	4609
29	4624	4639	4654	4669	4683	4698	4713	4728	4742	4757
30	4771	4786	4800	4814	4829	4843	4857	4871	4886	4900
31	4914	4928	4942	4955	4969	4983	4997	5011	5024	5038
32	5051	5065	5079	5092	5105	5119	5132	5145	5159	5172
33	5185	5198	5211	5224	5237	5250	5263	5276	5289	5302
34	5315	5328	5340	5353	5366	5378	5391	5403	5416	5428
35	5441	5453	5465	5478	5490	5502	5514	5527	5539	5551
36	5563	5575	5587	5599	5611	5623	5635	5647	5658	5670
37	5682	5694	5705	5717	5729	5740	5752	5763	5775	5786
38	5798	5809	5821	5832	5843	5855	5866	5877	5888	5899
39	5911	5922	5933	5944	5955	5966	5977	5988	5999	6010
40	6021	6031	6042	6053	6064	6075	6085	6096	6107	6117
41	6128	6138	6149	6160	6170	6180	6191	6201	6212	6222
42	6232	6243	6253	6263	6274	6284	6294	6304	6314	6325
43	6335	6345	6355	6365	6375	6385	6395	6405	6415	6425
44	6435	6444	6454	6464	6474	6484	6493	6503	6513	6522
45	6532	6542	6551	6561	6571	6580	6590	6599	6609	6618
46	6628	6637	6646	6656	6665	6675	6684	6693	6702	6712
47	6721	6730	6739	6749	6758	6767	6776	6785	6794	6803
48	6812	6821	6830	6839	6848	6857	6866	6875	6884	6893
49	6902	6911	6920	6928	6937	6946	6955	6964	6972	6981
50	6990	6998	7007	7016	7024	7033	7042	7050	7059	7067
51	7076	7084	7093	7101	7110	7118	7126	7135	7143	7152
52	7160	7168	7177	7185	7193	7202	7210	7218	7226	7235
53	7243	7251	7259	7267	7275	7284	7292	7300	7308	7316
54	7324	7332	7340	7348	7356	7364	7372	7380	738ε	7396
N.	0	1	2	3	4	5	6	7	8	9

Four Place Common Logarithms

N.	0	1	2	3	4	5	6	7	8	9
55	7404	7412	7419	7427	7435	7443	7451	7459	7466	7474
56	7482	7490	7497	7505	7513	7520	7528	7536	7543	7551
57	7559	7566	7574	7582	7589	7597	7604	7612	7619	7627
58	7634	7642	7649	7657	7664	7672	7679	7686	7694	7701
59	7709	7716	7723	7731	7738	7745	7752	7760	7767	7774
60	7782	7789	7796	7803	7810	7818	7825	7832	7839	7846
61	7853	7860	7868	7875	7882	7889	7896	7903	7910	7917
62	7924	7931	7938	7945	7952	7959	7966	7973	7980	7987
63	7993	8000	8007	8014	8021	8028	8035	8041	8048	8055
64	8062	8069	8075	8082	8089	8096	8102	8109	8118	8122
65	8129	8136	8142	8149	8156	8162	8169	8176	8182	8189
66	8195	8202	8209	8215	8222	8228	8235	8241	8248	8254
67	8261	8267	8274	8280	8287	8293	8299	8306	8312	8319
68	8325	8331	8338	8344	8351	8357	8363	8370	8386	8382
69	8388	8395	8401	8407	8414	8420	8426	8432	8439	8445
70	8451	8457	8463	8470	8476	8482	8488	8494	8500	8506
71	8513	8519	8525	8531	8537	8543	8549	8555	8561	8567
72	8573	8579	8585	8591	8597	8603	8603	8615	8621	8627
73	8633	8639	8645	8651	8657	8663	8669	8675	8681	8686
74	8692	8698	8704	8710	8716	8722	8727	8733	8739	8745
75	8751	8756	8762	8768	8774	8779	8785	8791	8797	8802
76	8808	8814	8820	8825	8831	8837	8842	8848	8854	8859
77	8865	8871	8876	8882	8887	8893	8899	8904	8910	8915
78	8921	8927	8932	8938	8943	8949	8954	8960	8965	8971
79	8976	8982	8987	8993	8998	9004	9009	9015	9020	9025
80	9031	9036	9042	9047	9053	9058	9063	9069	9074	9079
81	9085	9090	9096	9101	9106	9112	9117	9122	9128	9133
82	9138	9143	9149	9154	9159	9165	9170	9175	9180	9186
83	9191	9196	9201	9206	9212	9217	9222	9227	9232	9238
84	9243	9248	9253	9258	9263	9269	9274	9279	9284	9289
85	9294	9299	9304	9309	9315	9320	9325	9330	9335	9340
86	9345	9350	9355	9360	9365	9370	9375	9380	9385	9390
87	9395	9400	9405	9410	9415	9420	9425	9430	9435	9440
88	9445	9450	9455	9460	9465	9469	9474	9479	9484	9489
89	9494	9499	9504	9509	9513	9518	9523	9528	9533	9538
90	9542	9547	9552	9557	9562	9566	9571	9576	9581	9586
91	9590	9595	9600	9605	9609	9614	9619	9624	9628	9633
92	9638	9643	9647	9653	9657	9661	9666	9671	9675	9680
93	9685	9689	9694	9699	9703	9708	9713	9717	9722	9727
94	9731	9736	9741	9745	9750	9754	9759	9763	9768	9773
95	9777	9782	9786	9791	9795	9800	9805	9809	9814	9818
96	9823	9827	9832	9836	9841	9845	9850	9854	9859	9863
97	9868	9872	9877	9881	9886	9890	9894	9899	9903	9908
98	9912	9917	9921	9926	9930	9934	9939	9943	9948	9952
99	9956	9961	9965	9969	9974	9978	9983	9987	9991	9996
N.	0	1	2	3	4	5	6	7	8	9

Source: Appendix A reproduced from U.S. Air Force Manual 101–8, pp. 549–50.

Appendix B: Statistical Probabilities

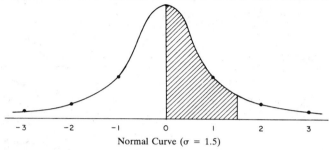

Normal Curve ($\sigma = 1.5$)

Areas under the Normal Curve

σ	.00	.01	.02	.03	.04	.05	.06	.07	.08	.09
0.0	.0000	.0040	.0080	.0120	.0160	.0199	.0239	.0279	.0319	.0359
0.1	.0398	.0438	.0478	.0517	.0557	.0596	.0636	.0675	.0714	.0753
0.2	.0793	.0832	.0871	.0910	.0948	.0987	.1026	.1064	.1103	.1141
0.3	.1179	.1217	.1255	.1293	.1331	.1368	.1406	.1443	.1480	.1517
0.4	.1554	.1591	.1628	.1664	.1700	.1736	.1772	.1808	.1844	.1879
0.5	.1915	.1950	.1985	.2019	.2054	.2098	.2123	.2157	.2190	.2224
0.6	.2257	.2291	.2324	.2357	.2389	.2422	.2454	.2486	.2518	.2549
0.7	.2580	.2612	.2642	.2673	.2704	.2734	.2764	.2794	.2823	.2852
0.8	.2881	.2910	.2939	.2967	.2995	.3023	.3051	.3078	.3106	.3133
0.9	.3159	.3186	.3212	.3238	.3264	.3289	.3315	.3340	.3365	.3389
1.0	.3413	.3438	.3461	.3485	.3508	.3531	.3554	.3577	.3599	.3621
1.1	.3643	.3665	.3686	.3708	.3729	.3749	.3770	.3790	.3810	.3830
1.2	.3849	.3869	.3888	.3907	.3925	.3944	.3962	.3980	.3997	.4015
1.3	.4032	.4049	.4066	.4082	.4099	.4115	.4131	.4147	.4162	.4177
1.4	.4192	.4207	.4222	.4236	.4251	.4265	.4279	.4292	.4306	.4319
1.5	.4332	.4345	.4357	.4370	.4382	.4394	.4406	.4418	.4429	.4441
1.6	.4452	.4463	.4474	.4484	.4495	.4505	.4515	.4525	.4535	.4545
1.7	.4554	.4564	.4573	.4582	.4591	.4599	.4608	.4616	.4625	.4633
1.8	.4641	.4649	.4656	.4664	.4671	.4678	.4686	.4693	.4699	.4706
1.9	.4713	.4719	.4726	.4732	.4738	.4744	.4750	.4756	.4761	.4767
2.0	.4772	.4778	.4783	.4788	.4793	.4798	.4803	.4808	.4812	.4817
2.1	.4821	.4826	.4830	.4834	.4838	.4842	.4846	.4850	.4854	.4857
2.2	.4861	.4864	.4868	.4871	.4875	.4878	.4881	.4884	.4887	.4890
2.3	.4893	.4896	.4898	.4901	.4904	.4906	.4909	.4911	.4913	.4916
2.4	.4918	.4920	.4922	.4925	.4927	.4929	.4931	.4932	.4934	.4936
2.5	.4938	.4940	.4941	.4943	.4945	.4946	.4948	.4949	.4951	.4952
2.6	.4953	.4955	.4956	.4957	.4959	.4960	.4961	.4962	.4963	.4964
2.7	.4965	.4966	.4967	.4968	.4969	.4970	.4971	.4972	.4973	.4974
2.8	.4974	.4975	.4976	.4977	.4977	.4978	.4979	.4979	.4980	.4981
2.9	.4981	.4982	.4982	.4983	.4984	.4984	.4985	.4985	.4986	.4986
3.0	.4986	.4987	.4987	.4988	.4988	.4989	.4989	.4989	.4990	.4990
3.1	.4990	.4991	.4991	.4991	.4992	.4992	.4992	.4992	.4993	.4993
3.2	.4993	.4993	.4994	.4994	.4994	.4994	.4994	.4995	.4995	.4995
3.3	.4995	.4995	.4995	.4996	.4996	.4996	.4996	.4996	.4996	.4997
3.4	.4997	.4997	.4997	.4997	.4997	.4997	.4997	.4997	.4998	.4998
3.5	.4998	.4998	.4998	.4998	.4998	.4998	.4998	.4998	.4998	.4998
3.6	.4998	.4998	.4999	.4999	.4999	.4999	.4999	.4999	.4999	.4999
3.7	.4999	.4999	.4999	.4999	.4999	.4999	.4999	.4999	.4999	.4999
3.8	.4999	.4999	.4999	.4999	.4999	.4999	.4999	.5000	.5000	.5000
3.9	.5000	.5000	.5000	.5000	.5000	.5000	.5000	.5000	.5000	.5000

Source: Appendix B reproduced from Mathematics, Vol. 3, NavPers 10073–A Rate Training Manual, pp. 58–59.

Appendix C: NBS Time and Frequency Standards

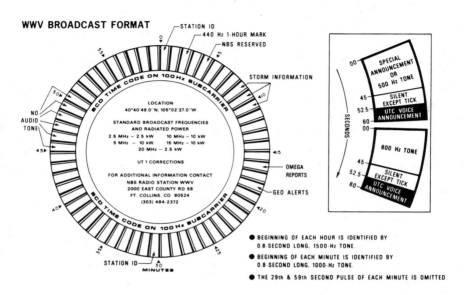

Source: Appendix C reproduced from NBS Special Publication 432, September 1979, pp. 1–3.

RADIO STATIONS WWV AND WWVH

NBS broadcasts continuous signals from its high-frequency radio stations WWV and WWVH. The radio frequencies used are 2.5, 5, 10, and 15 MHz. WWV also broadcasts on an additional frequency of 20 MHz. All frequencies carry the same program, but because of changes in ionospheric conditions, which sometimes adversely affect the signal transmissions, most receivers are not able to pick up the signal on all frequencies at all times in all locations. Except during times of severe magnetic disturbances, however—which make all radio transmissions almost impossible—listeners should be able to receive the signal on at least one of the broadcast frequencies. As a general rule, frequencies above 10 MHz provide the best daytime reception while the lower frequencies are best for nighttime reception. Services provided by these stations include time announcements, standard time intervals, standard frequencies, geophysical alerts, marine storm warnings, Omega Navigation System status reports, UTl time corrections, and BCD time code.

ACCURACY AND STABILITY

The time and frequency broadcasts are controlled by the primary NBS Frequency Standard in Boulder, Colorado. The frequencies as transmitted are accurate to within one part in 100 billion (1×10^{-11}) at all times. Deviations are normally less than one part in 1000 billion (1×10^{-12}) from day to day. However, changes in the propagation medium (causing Doppler effect, diurnal shifts, etc.) result in fluctuations in the carrier frequencies as received by the user that may be very much greater than the uncertainty described above.

At both WWV and WWVH, double sideband amplitude modulation is employed with 50 percent modulation on the steady tones, 25 percent for the BCD time code, 100 percent for seconds pulses, and 75 percent for voice.

TIME ANNOUNCEMENTS

Voice announcements are made from WWV and WWVH once every minute. To avoid confusion, a man's voice is used on WWV and a

woman's voice on WWVH. The WWVH announcement occurs first—
at 15 seconds before the minute—while the WWV announcement
occurs at 7 1/2 seconds before the minute. Though the announcements occur at different times, the tone markers referred to are
transmitted simultaneously from both stations. However, they may
not be received at the same time due to propagation effects.

   The time referred to in the announcements is "Coordinated
Universal Time" (UTC). It is coordinated through international
agreements by the International Time Bureau (BIH) so that time
signals broadcast from the many stations such as WWV throughout
the world will be in close agreement.

   The specific hour and minute mentioned is actually the time at
the time zone centered around Greenwich, England, and may be
considered generally equivalent to the more well-known "Greenwich
Mean Time" (GMT). UTC time differs from your local time only
by an integral number of hours. By knowing your own local time
zone and using a chart of world time zones, the appropriate number
of hours to add or subtract from UTC to obtain local time can be
determined. The UTC time announcements are expressed in the 24-
hour clock system—i.e., the hours are numbered beginning with 00
hours at midnight through 12 hours at noon to 23 hours, 59 minutes
just before the next midnight.

STANDARD TIME INTERVALS

The most frequent sounds heard on WWV and WWVH are the pulses
that mark the seconds of each minute, except for the 29th and 59th
pulses which are omitted completely. The first pulse of every hour
is an 800-millisecond pulse of 1500 Hz. The first pulse of every minute is an 800-millisecond pulse of 1000 Hz at WWV and 1200 Hz at
WWVH. The remaining seconds pulses are brief audio bursts (5-
millisecond pulses of 1000 Hz at WWV and 1200 Hz at WWVH) that
resemble the ticking of a clock. All pulses commence at the beginning of each second. They are given by means of double-sideband
amplitude modulation.

   Each seconds pulse is preceded by 10 milliseconds of silence
and followed by 25 milliseconds of silence to avoid interference which
might make it difficult or impossible to pick out the seconds pulses.
Thus, there is a 40-millisecond protected zone around each seconds
pulse.

STANDARD AUDIO FREQUENCIES

In alternate minutes during most of each hour, 500 or 600 Hz audio tones are broadcast. A 440 Hz tone, the musical note A above middle C, is broadcast once each hour. In addition to being a musical standard, the 440 Hz tone can be used to provide an hourly marker for chart recorders or other automated devices.

Appendix D: Military Specifications

The following documents are included to show the student the range and detail of typical military specifications for the procurement of instruments and services. In addition to becoming familiar with these specifications, the student can also use them to compare the capabilities that the government as a buyer requires compared to what the seller usually defines.

Multimeter, Electronic

This specification is approved for use by the US Army Missile Research and Development Command, Department of the Army, and is available for use by all Departments and Agencies of the Department of Defense.

1. SCOPE

1.1 Objective. The objective of this specification is delineate of high quality, wide range, precise, reliable, and rugged electronic multimeter that will meet the requirements of a current major missile system. It shall have versatile application, must meet the rigors of environmental and field conditions without drift or deterioration of accuracy, reliability, or stability, and must meet rigorous space limitations.

1.2 Description. The electronic multimeter, hereafter called multimeter, shall combine into one instrument and its accessories the capabilities necessary for a multitude of applications, all of which are necessary for effectively maintaining a current major missile system. The multimeter shall be capable of being used as a portable instrument, self-contained, and also as a panel assembly. Mounting in the various panels shall be by suitable adapters.

2. APPLICABLE DOCUMENTS

2.1 The following documents of the issue in effect on date of invitation for bids or request for proposal, form a part of this specification to the extent specified herein.

Source: Appendix D reproduced from U.S. Government Printing Office documents 1979–603–022/1979 and 1979–603–022/1762.

SPECIFICATIONS

Military

MIL-E-1	Electronic Tubes, General Specification for
MIL-T-27	Transformers and Inductors (Audio, Power and High Power Pulse) General Specification for
MIL-C-3432	Cable and Wire, Electrical (Power and Control, Flexible and Extra Flexible, 300 and 600 Volts)
MIL-C-3767	Connectors, Plug and Receptacle, (Power, Bladed Type) General Specification for
MIL-M-10304	Meters, Electrical Indicating, Panel Type, Ruggedized, General Specification for
MIL-E-15090	Enamel, Equipment, Light Gray (Formula No. 111)
MIL-S-19500	Semiconductor Devices, General Specification for

STANDARDS

Military

MIL-STD-130	Identification Marking of US Military Property
MIL-STD-202	Test Methods for Electronics and Electrical Component Parts
MIL-STD-461	Electromagnetic Interference Characteristics Requirements for Equipment
MIL-STD-785	Reliability Program for Systms and Equipment Development and Production
MIL-STD-750	Test Methods for Semiconductor Devices
MIL-STD-810	Environmental Test Methods
MIL-STD-1188	Commercial Packaging of Supplies and Equipment
MS 3102	Connector, Receptacle, Electrical, Box Mounting
MS 3106	Connector, Plug, Electric, Straight

(Copies of specifications, standards, drawings, and publications required by contractors in connection with specific procurement functions should be obtained from the procuring activity or as directed by the contracting officer.)

3. REQUIREMENTS

3.1 _Construction._ The multimeter shall meet the environmental, physical, maintenance, and functional requirements delineated herein and in accordance with the appropriate portions of the documents in Section 2.

3.2 _Physical characteristics._ The multimeter shall consist of a functional assembly, case and over assembly, and accessories and shall be used both as a portable instrument and as a panel assembly.

3.2.1 <u>Functional assembly</u>. The functional assembly shall constitute the complete functional portion of the multimeter less the accessories and the case and cover assembly, and shall have the following physical characteristics:

a. The functional assembly shall have a front panel, the dimensions of which are 10 7/8 in. wide x 6 5/8 in high. The panel thickness shall be 3/16 in.

b. Without modification, the multimeter shall be easily converted into a second configuration with all features remaining the same except the front panel, which now shall be 6 5/8 in. wide x 10 7/8 in. high.

c. The front panel shall be corrosion resistant metal, and all markings thereon shall be etched, engraved, embossed, or otherwise permanently affixed so as not to deteriorate under use.

d. Shielding shall be included as part of the functional assembly, for the purpose of protecting the electrical portion of the functional assembly. This shielding shall be secured to the front panel and shall serve as a dust cover when the multimeter is used as a panel assembly. The maximum outside dimensions of the shielding shall be 9 1/4 inches x 5 5/16 inches x 4 5/16 inches deep; no bolts, fasteners, or other means of securing the shielding to the front panel shall extend outside the shielding dimensions behind the front panel.

e. The location of the shielding on the front panel shall conform to the following:

(1) For the 10 7/8 inch wide x 6 5/8 inch high front panel, the outer edge of the shielding shall be located 21/32 inch from the top edge of the panel and 21/32 inch from the bottom edge of the panel.

(2) For the 6 5/8 inch wide x 10 7/8 inch high front panel, the outer edge of the shielding shall be located 1 inch from the top edge of the panel and 5/8 inch from the bottom edge of the panel.

f. The maximum useable front panel space shall be the area confined by the shielding (9 1/4 inches x 5 5/16 inches).

g. The shielding shall be readily removable from the front panel for maintenance and servicing of the multimeter.

h. All connections to the multimeter shall be made on the front panel within the bounds of the useable panel space (<u>f</u> above); however, a second AC power connection shall be provided in the rear of the shielding (for use when the multimeter is used as a panel assembly). These two input power connectors (one in the front panel and one in the rear shielding) shall be MS3102R14S-7P Connectors per MS3102.

i. The input power cable shall be Type CO-03MGF(3/16)0365 per MIL-C-3432. Electrical connector Type UP12M per MIL-C-3767/4 shall be connected to one end of the cable; connector MS3106R14S-7S per MS3106 shall be connected to the other end.

j. Knobs and pointers shall be securely fastened to the shafts they control. They shall be made of a durable material and shall be of a size and shape to facilitate rotation under adverse operating conditions. All knob and pointer markings shall be engraved, etched, embossed, or otherwise permanently affixed so that they will not be obliterated by constant use. All knobs and pointers shall be readily removable without special tools.

k. The controls for the multimeter shall include but not be limited to the following:

    (1) range switch

    (2) damping

    (3) DC polarity reversing

    (4) power switch

    (5) zero adjustment, external

    (6) function (volts, ohms, hi, low, etc.)

NOTE: Combination of functions shall be permitted provided flexibility of use is not impaired.

    1. Control shafts shall be flatted or otherwise provided with means for accurately reinstalling knobs, pointers, and dials, subsequent to servicing the multimeter.

    m. The indicating meter shall meet the following criteria:

    (1) The indicating meter shall have torsional suspension with no backlash.

    (2) The mechanical zero adjustment may be located on the rear of the meter.

    (3) The meter dimensions shall not be less than 4 1/4 inches square; the meter scale length shall not be less than 3 1/2 inches.

    (4) The meter shall have a mirrored dial, a knife edge pointer, and a white background dial. The white dial shall have black markings. The pointer shall be black over the white portions of the dial and shall be white over the mirrored portion of the dial.

    (5) The meter glass shall be static-free and resistant to fracture due to minor shock or impact. It shall be unaffected as to color and transparency as a result of mild abrasion, full continuous sunlight, and other environmental operating conditions.

    (6) The meter shall have means for full dial illumination, without using external lighting beyond the shielding (which includes the front panel facing).

    (7) The meter shall meet the requirements of MIL-M-10304 not specifically covered herein.

    n. The multimeter shall be operable in any physical position, with no effect on the operation or accuracy.

3.2.2 _Portability_. For use as a portable instrument, the multimeter shall have a case and cover assembly, which shall conform to the following:

    a. For portable application, the electrical portion of the functional assembly shall be operable within the case assembly. The case assembly shall be secured to the front panel (it is acceptable to drill holes in the front panel, outside the shielding, for this purpose). The cover assembly shall be secured to the case assembly. The overall maximum dimensions of the multimeter (case and cover assembly included) shall be 11 1/2 inches x 7 1/4 inches x 6 1/4 inches. All fastenings or other means of securing the case and cover assembly shall not protude beyond these dimensions.

    b. The cover assembly shall be removable from the case assembly without the use of tools. The case assembly shall be readily removable from the functional assembly with simple tools.

c. The case and cover assembly shall be non-conducting, resistant to high impact, and shall fully protect the functional assembly from the elements.

d. No cables, connectors, or leads shall pass through any portion of the case and cover assembly.

e. A folding carrying handle shall be provided on the case, recessed to facilitate stacking in storage.

f. Feet shall be provided on the back and the bottom of the case assembly, so that the multimeter with the cover removed, may be utilized in either the vertical or horizontal position. (For the vertical position, the front panel shall be 10 7/8 inches high x 6 5/8 inches wide.) These feet shall not exceed the maximum allowable dimensions (a above), and shall preferably be integral with the case. They shall not slide readily on smooth surfaces or mar such surfaces easily.

g. The multimeter shall have a front panel adapter (or adepters) which shall extend the front panel dimensions to 19 inches wide x 7 inches high. (The case assembly will not be connected to the front panel when this adapter is used.) It is desirable (but not required) for this adapter (or adapters) to fit flush with the existing front panel. Any suitable means may be used for fastening this adapter (or adapters) to the panel; however, the adapter (or adapters) shall readily be installed and removed with simple tools. (Also, the panel adapter junction shall be rigid· and shall withstand all shock and vibration tests specified herein and according to the applicable documents in Section 2). If holes were drilled in the front panel (a above), they may be used. With the adapter (or adapters) attached, the maximum depth from the front adapter (or adapters) facing to the rear of the shielding shall be 4 1/2 inches. The material and finish of the adapter (or adapters) shall be the same as that of the front panel itself. The multimeter front panel shall be centered horizontally within the adapter; that is, the width of the panel adapter shall be 4 1/16 inch on each side of the panel. The 3/8 inch strip of adapter material (necessary for obtaining the 7 inch high dimension) shall be located along the bottom of the panel. For mounting purposes, four notches shall be cut out of the panel adapter edges. Considering the 19 inch wide x 7 inch high front surface, the centers of two notches are located 1 1/2 inches from the top adapter edge, down both sides. The centers of the other two notches are located 4 inches below the centers of the top notches on both sides. The widths of all four notches at the adapter edges are 1/4 inch. A 1/4 inch slot then extends inward for 3/8 inch; then at this point the slot "rounds" with a 1/8 inch radius, making the length of the notch 1/2 inch. These measurements are typical at all four notches.

3.2.3 _Accessories_. The accessories shall include the following as required:

a. RF probes with leads, minimum length 5 ft.

b. DC probes with leads, minimum length 5 ft.

c. Grounding and straight through test leads, minimum length 5 ft.

d. Power cable, minimum length 8 ft.

e. Adapter connectors

3.2.3.1 All cables and test leads shall be detachable.

3.2.4 _Materials_. Materials selected shall meet all the electrical, environmental, and functional requirements specified herein. Primary consideration shall be given to the selection of materials not subject to attack by fungus or high humidity. Only a minimum of fungus proofing material shall be applied in accordance with MIL-STD-202 snd shall be applied only to materials subject to attack by fungus. Where fungus proofing is required, it shall be applied to the materials prior to assembly into the multimeter. An inorganic vehicle shall be used to apply the fungus proofing material, which is not in itself subject to attack by fungus or high humidity. Standard materials shall be used throughout, whenever practicable. Unless otherwise indicated herein, all piece parts shall be militarized types to the maximum extent practicable. Proprietary items shall be held to a minimum.

3.2.4.1 <u>Finish</u>. Finish of the case and cover assembly shall be that of the material used. The color shall preferably be grey; however, green may be used as an alternate. Finish of the panel shall be grey, enamel, Type III, Class 2, per MIL-E-15090.

3.2.5 <u>Operating and nonoperating condtions</u>. Operating and nonoperating conditions delineated in MIL-STD-202 and MIL-STD-810 shall be met except as indicated herein.

3.2.5.1 <u>Shock and vibration</u>. Shock and vibration resistance capability shall include transportability in military vehicles over rough roads without damage or deterioration in performance.

3.2.6 <u>Construction</u>. Packaging of the multimeter shall be such that repairs may be made in the field; modular construction shall be employed.

3.3 <u>Electrical characteristics</u>.

3.3.1 <u>Functional assembly</u>. The functional assembly shall have the following electrical characteristics.

    a. The DC voltage ranges shall be 2.5mV, 5mV, 10mV, 20mV, 50mV, 0.1V, 0.25V, 0.50V, 1.0V, 2.5V, 5.0V, 10.0V, 25V, 50V, 100V, 250V, 500V, and 1000V full scale.

    b. The AC voltage ranges shall be as follows:

        (1) 2.5mV, 5.0mV, 10mV, 25mV, 50mV, 0.1V, 0.25V, 0.50V, 1.0V, 2.5V, 5.0V, 10.0V, 25V, 50V, 100V, 250V, 500V, and 1000V RMS full scale within the frequency range from 3 cps to 150 kc.

        (2) 2.5mV, 5.0mV, 10mV, 25mV, 50mV, 0.1V, 0.25V, 0.5V, 1.0V, 2.5V, 5.0V, 10V, 25V, 50V, 100V, and 250V RMS full scale within the frequency range from 150 kc to 250mc.

    c. The voltage accuracy shall be within $\pm$ 2% of the scale reading for all ranges, both AC and DC.

    d. A decibel scale shall be provided with a range from -20 to +62 DB. There shall be no less than 5 ranges. The reference shall be 1 milliwat into 600 ohms at zero BD.

    e. Eight resistance ranges shall be provided: 10 ohms, 100 ohms, 1K, 10K, 100K, 1 megohm, 10 megohm, and 100 megohm center scale. The accuracy of all resistance measurements shall be within $\pm$ 2% of arc length or within $\pm$ 2 degrees of arc, whichever is smaller.

    f. The input impedance shall meet the following criteria:

        (1) From dc to 150 kc, no less than 100 megohms resistive, with a distributed shunt capacitance not to exceed 10 picofarad (pf).

        (2) Open grid input shall be provided for the 0-500 vdc ranges. This input shall be no less than 5000 megohms resistive with a shunt distributed capacitance not to exceed 5 pf.

(3) From 150 kc to 100 mc, no less than 10 megohms resistive, with a distributed shunt capacitance not to exceed 2.5 pf.

g. The multimeter shall operate from a power source of 115V ± 10 vdc at frequencies of 50 to 440 cps.

3.3.2 Grounding. Grounding shall meet the following requirements:

a. The front panel and the shielding shall be connected to the green wire in the power cable, via the power input receptacle.

b. The internal circuitry shall be isolated from the front panel and the external shielding.

c. Two binding posts, universal, shall be provided on the front panel. One of these shall be connected to the front panel, and the other shall be connected to the low potential junction of the electronic curcuitry of the multimeter. A connector adapter shall be proviced such that these two binding posts may be optionally connected together.

d. Shielding of probes shall be such that no potential shall exist between the exposed probe shields and connectors, and the multimeter panel.

3.3.3 Above ground potential. Capability shall be inherent to measure potentials between any two points above ground without creating a safety hazard and without damage to the multimeter or unit under test. This shall apply to any range and to both AC and DC potentials.

3.3.4 DC probe. The DC probe shall be fitted so that High Voltage Probe MX 2517/U, per Drawing 9113325, may be used. This will extend the range of the multimeter to 30,000 volts DC at a current drain not exceeding 2.5 microamperes.

3.3.5 Stability. Stability of the multimeter shall meet the following criteria:

a. No harmonic or subharmonic of the power input shall have any effect on the operation of the multimeter.

b. Changing ranges shall not affect the zero or full scale readings. Rezeroing shall not be required.

c. Environmental conditions, either in service or in storage, shall not have a combined effect on accuracy of more than ± 1%.

d. Drift in performance shall not be greater than 1%.

(1) One hour after the initial warm-up period of 20 minutes (tube circuit), or

(2) 20 minutes after the initial warm-up period of 2 minutes (solid state circuit).

e. Response time shall not be greater than 2 seconds for DC and above 10 cps, and not greater than 3 seconds for 3 cps to 10 cps.

f. A damping control shall be provided to stabilize readings from 3 to 10 cps. This need not be a separate control and may be switched off in order that readings shall not be sluggish.

g. The equipment shall be designed to give optimum performance and reliable service during continuous or intermittent operating periods of at least 1000 hours under the conditions specified here without the necessity for major adjustment, calibration, or service.

3.3.6 <u>Safety feature</u>. A safety feature which shall in no way cause damage to the unit under test shall be inherent to the multimeter to fully protect the meter and electrical circuitry from any accidental instantaneous or prolonged overload on any range (AC and DC). A readily accessible reset shall be available.

3.4 <u>Calibration adjustments</u>. Calibration adjustments of the multimeter shall meet the following criteria:

      a. Provision shall be provided for self-calibration checking of the multimeter at any time without recourse to changing scales or range settings and without removing the test prods from the unit under test.

      b. An external control shall be provided on the front panel for zero voltage adjustment; the zero adjustment shall not change when the voltage range switch is operated. One adjustment shall be sufficient for all AC and DC voltage ranges.

      c. If required, separate internal AC and DC zero voltage range adjustments shall be provided to make the front panel zero adjustment coincident on AC and DC ranges. These internal adjustments shall be accessible through the rear of the shielding (but not the case assembly).

      d. Calibration adjustments other than those specified shall not be accessible externally. If required to maintain accuracy, additional calibration adjustments (accessible only upon removal of the shielding) shall be provided to compensate for aging and tube or solid state electronic device replacement. Recalibration shall not normally be required as a result of changing a tube or solid state device.

      e. Interaction of controls or calibration adjustments shall be not less than 1000:1.

3.5 <u>Environmental</u>.

3.5.1 <u>Operating conditions</u>. The multimeter shall be capable of meeting its operating requirements without recalibration after being subjected to environmental conditions specified herein. After a warm-up period, the multimeter shall operate satisfactorily under any possible combination of the following ambient conditions:

      a. Temperature ranging from minus 51°C to plus 85°C for intermittent and continuous periods.

      b. Relative humidities ranging up to 95% for both intermittent and continuous periods.

      c. Barometric pressures ranging from to 20.6 inches of mercury (approximating an altitude of 10,000 feet).

3.5.2 <u>Nonoperating conditions</u>. The multimeter shall not be damaged nor shall its performance be impaired from storage in any position or when subjected to any possible combination of the following:

      a. Temperature ranging from minus 51°C to plus 85°C for extended periods.

      b. Relative humidity ranging up to 100% for extended periods.

      c. Barometric pressure ranging down to 3.4 inches of mercury (approximating an altitude of 50,000 feet).

      d. Fungus and fungus supporting conditions for extended periods.

3.6 <u>Instruction manual</u>. Two copies of an operating instruction manual shall accompany each multimeter. This manual shall include schematic diagrams and all necessary maintenance procedures.

3.7 <u>Spare tube kit</u>. One complete set of spare tubes (or solid state devices) shall be furnished with the multimeter.

3.8 <u>Schematic diagram</u>. A durable plastic schematic diagram of the multimeter shall be secured to the inside of the shelding. "Where applicable, waveforms shall appear at test points on the schematic."

3.9 <u>Marking</u>. Marking for identification shall conform to the requirements specified in MIL-STD-130.

3.10 <u>Electron Tubes and semiconductors</u>. All electron tubes and semiconductors used in the multimeter shall conform to requirements specified in MIL-E-1 or MIL-S-19500, whichever is applicable. All tests on semiconductors shall conform to MIL-STD-750.

3.11 <u>Transformers</u>. All transformers and inductors used in the multimeter shall conform to the requirements specified in MIL-T-27, and shall have a life expectancy X, grade 4 or 5.

3.12 <u>Interference suppression</u>. The multimeter shall conform to the requirements of MIL-STD-461 and the interference limits set forth herein.

3.13 <u>Workmanship</u>. All components shall be fitted properly and firmly into their respective positions. All mechanical joints and electrical connections shall be mechanically secure and electrically sound. All controls shall operate freely, without slippage, sticking, or binding. Electrical connections shall be flexible between shock mounted components and their foundations. Surfaces of glazed ceramic components shall be smooth, uniform, and free from cracks. There shall be no evidence of corrosion, burrs, roughness, sharp edges, or other defects, which would adversely affect its use, or suitability for the purpose intended.

4. QUALITY ASSURANCE PROVISIONS

4.1 <u>Responsibility for inspection</u>. Unless otherwise specified in the contract, the contractor is responsible for the performance of all inspection requirements as specified herein. Except as otherwise specified in the contract, the contractor may use his own or any other facilities suitable for the performance of the inspection requirements specified herein, unless disapproved by the Government. The Government reserves the right to perform any of the inspections set forth in the specification where such inspections are deemed necessary to assure supplies and services conform to prescribed requirements.

4.2 <u>Preproduction sample</u>. Unless otherwise specified (see 6.2), one unit, which has been manufactured under the same conditions as those proposed for subsequent unit of production, shall be submitted for Government approval. The unit shall be subjected to all examinations and tests specified herein. Unless otherwise specified (see 6.2), the Government will perform the examination and tests at the contractor's plant. If the preproduction sample does not meet all requirements of this specification, it will be rejected and returned to the contractor.

4.3 <u>Inspection</u>. Unless otherwise specified (see 6.2) by the procuring activity, acceptance inspection for production shall be 100 percent, with the exception of the environmental tests which shall be performed on the preproduction sample only.

4.3.1 <u>Lot size</u>. The inspection lot size shall not exceed that quantity of items produced in one continuous operation by the same manufacturer and manufactured in accordance with the same drawing, drawing revision, specification and specification revision.

4.4 Testing.

4.4.1 Test method. "Apply signals specified in paragraphs 3.3.1.a and 3.3.1.b as measured on a laboratory standard instrument having an accuracy within plus or minus 0.25 percent at the value being measured". For each voltage range, obtain zero voltage, one quarter scale voltage, one half scale voltage, three quarter scale voltage, and full scale voltage, and measure each of these voltages on the multimeter under test. Measure the resistance of resistors having an accuracy within ± 0.25 percent. Measure at least three resistance values, one of which is equal to approximately two-thirds of the meter deflection full scale arc, for each range specified (see 3.3.1.e).

4.4.2 Accuracy test. The multimeter shall be tested for accuracy of calibration at all measurements specified in 4.4.1 at -65°C, +25°C, and +85°C, with relative humidities of zero percent, 40 percent, and 95 percent (at each possible combination). The percent of error shall not exceed that specified in 3.3.1.c and 3.3.1.e.

4.4.3 Input impedance. The input impedance of the multimeter shall conform to the requirements specified in 3.3.1.f.

4.4.4 Grounding. The grounding of the multimeter shall conform to the requirements specified in 3.3.2.

4.4.5 Input power. The multimeter shall operate from a power source as specified in 3.3.1.g.

4.4.6 Measuring potentials. The potential measuring capability of potentials above ground shall conform to the requirements specified in 3.3.3.

4.4.7 DC probe. The DC probe shall conform to the requirements specified in 3.3.4.

4.4.8 Stability. The stability of the multimeter shall conform to the requirements specified in 3.3.5.

4.4.9 Safety feature. A safety feature shall be inherent to the multimeter, and shall conform to the requirements specified in 3.3.6.

4.4.10 Calibration adjustments. The calibration adjustments of the mulitmeter shall conform to the requirements specified in 3.4.

4.4.11 Physical characteristics. The physical characteristics of the multimeter shall conform to the requirements specified in 3.2.

4.4.12 Reliability. Design for reliability shall occur simultaneously with, rather than separately from, the design to achieve electrical and mechanical characteristics. The multimeter shall be designed to obtain good consistent performance rather than occasional perfect performance. Reliability of the multimeter shall conform to the requirements specified herein and in MIL-STD-785.

4.5 Environmental.

4.5.1 Temperature, relative humidity and altitude. The mulitmeter shall be capable of withstanding without damage any possible combination of the ambient conditions specified in 3.5. MIL-STD-810, method 501, for high temperature; method 502, for low temperature; method 507 for humidity; and method 500 for low pressure shall be used to the extent applicable.

4.5.2 <u>Fungus</u>. The multimeter shall not be damaged nor its performance impaired when adjusted to the fungus spotting tests specified in MIL-STD-810, method 508, fungus (see 3.2.4 and 3.5.2d).

4.5.3 <u>Vibration</u>. The multimeter shall be capable of withstanding without damage the applicable test specified in MIL-STD-810, method 514, vibration (see 3.2.5.1).

4.5.4 <u>Shock</u>. The multimeter shall be capable of withstanding without damage the applicable test specified in MIL-STD-810, method 5.6, Shock (see 3.2.5.1).

4.6 <u>Preservation, packing and marking</u>. Preservation, packing, and marking shall be inspected to insure compliance with the provisions set forth in the packaging data sheet.

5. PACKAGING

5.1 <u>Preservation, packing, unitization, and marking</u> - Level A, B or commercial packaging (see 6.2).

5.1.1 <u>Level A or B.</u> Preservation, packing, unitization and marking shall be in accordance with packaging data sheet APN 9113325 and specifications referenced therein.

5.1.2 <u>Commercial</u>. Preservation, packing, unitization and marking shall be in accordance with MIL-STD-1188.

6. NOTES

6.1 <u>Intended use</u>. The electronic multimeter is intended for use in the general servicing of electrical and electronic equipment.

6.2 <u>Ordering data.</u> Procurement documents should specify the following:

    a. Title, number and date of this specification.

    b. Whether a preproduction sample is required.

    c. Where preproduction sample tests are to be performed.

    d. Inspection level, if other than specified.

    e. Selection of applicable levels of preservation, and packing.

    f. Applicable stock number, part numbers and official nomenclature.

6.3 <u>Changes from previous issue</u>. The margins of this specification are marked with an asterisk to indicate where changes (additions, modifications, corrections, deletions) from the previous issue were made. This was done as a conveniece only and the Government assumes no liability whatsoever for any inaccuracies in these notations. Bidders and contractors are cautioned to evaluate the requirements of this document based on the entire content irrespective of the marginal notations and relationship to the last previous issue.

Bridge, Impedance ZM-78/U

This specification is approved for use by the Communications Research and Development Command, Department of the Army, and is available for use by all Departments and Agencies of the Department of Defense.

1. SCOPE

1.1 <u>Scope</u>. This specification defines an impedance bridge capable of measuring resistance, capacitance, inductance, storage factor and dissipation factor and is hereinafter called the equipment.

1.2 <u>Classification</u>. The equipment defined by this specification shall be Type II, Class 5, Style E, Color R per MIL-T-28800 and as herein with the convertible/rack capability.

2. APPLICABLE DOCUMENTS

2.1 <u>Issues of documents</u>. The following documents of the issue in effect on the date of invitation for bids, or the request for proposal, form a part of this specification to the extent specified herein.

SPECIFICATIONS

MILITARY

MIL-T-28800	Test Equipment for Use with Electrical and Electronic Equipment, General Specification

STANDARDS

MILITARY

MIL-STD-461	Electromagnetic Interference Characteristics, Requirements
MIL-STD-462	Electromagnetic Interference Characteristics, Measurement
MIL-STD-781	Reliability Tests, Exponential Distribution

(Copies of specifications, standards, drawings and publications required by contractors in connection with specific procurement functions should be obtained from the procuring activity or as directed by the contracting officer).

3. REQUIREMENTS

3.1 <u>Classification of requirements</u>. The requirements for the equipment are classified as follows:

3.2 <u>Safety</u>. Unless otherwise specified herein, the equipment shall comply with the Type II safety requirements specified in MIL-T-28800.

3.3 <u>Parts, materials and processes</u>. Unless otherwise specified herein, the equipment shall comply with the Type II parts, materials and processes requirements of MIL-T-28800.

3.3.1 <u>Restricted materials</u>. Equipment shall comply with the restricted material requirements of MIL-T-28800 including the requirement for mercury or radioactive materials and shall contain no combination of materials which cause deterioration of any material contained in the equipment due to effects of outgassing.

3.4 <u>Design and construction</u>. Unless otherwise specified herein, the equipment shall comply with the Type II, Class 5, Style E, Color R, design and construction requirements of MIL-T-28800 with the convertible/rack-mountable capability.

3.4.1 <u>First article</u>. When specified, the contractor shall furnish five sample equipments for first article inspection and approval. (See 4.3 and 6.2)

3.4.2 <u>Mainframe plug-in concept</u>. Equipment utilizing externally accessible, externally removable subassemblies (mainframe plug-in concept) do not meet the requirements of this specification and therefore are unacceptable. Items using this concept may be offered, provided the plug-in(s) shall not be removable by access through the front panel or rear panel with the unaided hand. A system part number shall be assigned to any such mainframe plug-in combination required to meet the requirements of this specification. The system part number shall be marked on the mainframe identification plate and any plug-in(s).

3.4.3 <u>Controls</u>. Unless otherwise specified herein, built-in adjustments and compensating devices shall not be externally accessible.

3.4.3.1 <u>Front panel controls</u>. All controls which are required to operate the equipment throughout its specified performance characteristics, shall be located on the front panel.

3.4.4 <u>Accessibility</u>. The equipment shall be constructed so that:

3.4.4.1 <u>Subassemblies</u>. Subassemblies and chassis components can be removed without removing any other hard wired subassembly, printed circuit card or component.

3.4.4.2 <u>Adjustments</u>. Adjustments can be made without removing any component, printed circuit card or subassembly except the use of extender cards is permitted.

3.4.4.3 <u>Printed circuit cards</u>. Printed circuit cards can be removed without the need to unsolder cables and interconnecting wiring (connections to all printed circuit cards shall be through pin and socket connectors). Printed circuit cards (mother boards) designed primarily to distribute power and signals to other printed cards (daughter boards) are excluded from this requirement. When such mother boards are used, they shall be accessible from both sides to allow maintenance testing.

3.4.4.4 Indicator lights. Unless approved by the procuring activity upon presentation of acceptable reliability data, indicator lights other than light emitting diodes (LEDs) shall be accessible from the operator's side of the front panel.

3.4.4.5 Encapsulation and embedment. Encapsulation and embedment (potting) of subassemblies shall not be used.

3.4.5 Solid State construction. Unless otherwise specified herein, the equipment shall be of solid-state, modular, miniaturized construction.

3.5 Electrical power sources and connections. Unless otherwise specified herein, the equipment shall comply with the Type II electrical power sources and connections requirements of MIL-T-28800. The equipment shall operate from a nominal 115/230 volts, single phase, 50, 60 and 400Hz source.

3.5.1 Maximum power. The maximum power consumption of the equipment shall be 15 watts.

3.5.2 Input power selection device. An input power selection device shall be provided for selection of input power voltages of 115 VAC or 230 VAC. Provision shall be incorporated to prevent accidental switching. When the equipment is delivered, the power selection device shall be in the 115 VAC position.

3.5.3 Fuses and circuit breakers. Fuses and circuit breakers shall be in accordance with MIL-T-28800. (115VAC/230VAC) Either common or separate fuseholders may be provided. If only one fuseholder is used (common), the equipment shall be provided with the 115VAC fuse installed and the 230VAC fuse shall be stowed with the accessories.

3.5.4 Input power switch. A front panel mounted power switch shall be provided. The ON position shall have panel identification lights for AC operation. The switch shall break both sides of the power source.

3.6 Dimensions and weight.

3.6.1 Dimensions. The overall dimensions shall be 317.5mm (12.50 in) maximum height, 292.1mm (11.50 in) maximum depth and the width as specified in MIL-T-28800 for rack-mounted equipment. A blank plate may be required to satisfy incremental height requirements.

3.6.2 Weight. The maximum weight of the equipment shall be 17KG (37 lbs).

3.7 Enclosure requirements. Unless otherwise specified herein, the equipment shall comply with the Style E enclosure requirements of MIL-T-28800.

3.8 Marking and identification. Unless otherwise specified herein, the equipment shall comply with the Type II marking and identification requirements of MIL-T-28800.

3.8.1 Supplemental identification plate. The supplemental identification plate specified in MIL-T-28800 shall contain the following data only:

    a. Nomenclature.
    b. Procurement instrument identification number (PIIN).
    c. Serial number.
    d. National stock number.
    e. US

3.9 <u>Environmental requirements</u>. Unless otherwise specified herein, the equipment shall comply with the Class 5 environmental requirements of MIL-T-28800.

3.9.1 <u>Electromagnetic interference</u>. The equipment shall comply with the following emission and susceptibility requirements of Notice 4, MIL-STD-461.

CE02	CS02	RE02.1	RS03
CE04	CS06	RE02	

RE02.1 and RS03 shall be performed from 14 KHz to 1 GHz, with RS03 at a susceptibility level of one volt per meter (1V/m).

3.9.2 <u>Humidity</u>. The equipment shall meet humidity requirements in accordance with the test specified in paragraph 4.5.5.1.1.3 of MIL-T-28800.

3.9.3 <u>Vibration</u>. The equipment shall comply with the Class 5 vibration requirements of MIL-T-28800, except that the equipment need not be operating during vibration.

3.10 <u>Reliability requirements</u>.

3.10.1 <u>Reliability burn-in</u>. Each equipment delivered against this specification shall be submitted to a minimum 96-hour on-time burn-in procedure as specified in 4.4.4. The last 24 hours of burn-in shall be failure free.

3.10.2 <u>Reliability</u>. Reliability shall comply with requirements as specified herein. The specified MTBF shall be 3500 hours when tested as specified in 4.4.4. A failure shall be defined in MIL-STD-781, para 3.1, and as any departure from the required performance or operation of the required accuracies (not correctable by normal use of the operating controls) after the test is initiated. Test Level B, MIL-STD-781 shall be the required test level.

3.10.3 <u>Maintainability requirements</u>. The equipment shall comply with the Type II maintainability requirements of MIL-T-28800. (See 6.4)

3.11 <u>Performance characteristics</u>.

3.11.1 <u>Capacitance range</u>. The Capacitance Measurement capability shall extend from 1 picofarad to 1200 microfarad in not more than 8 ranges, with an accuracy of ± 0.2 percent.

3.11.1.1 <u>Dissipation factor</u>. The equipment shall have the capability to measure the dissipation factor within the range of 0.001 to 10 with an accuracy of +/- 5 percent.

3.11.2 <u>Inductance</u>. The inductance measurement capability shall extend from 1 microhenry to 1100 henry in not more than 8 ranges. The accuracy for each range shall not be less than +/-.1 percent.

3.11.2.1 <u>Internal signal source</u>. The equipment shall have an internal signal source of 1KHz +/- 3 percent with a voltage output continuously variable from 0 to 12 volts on open circuit.

3.11.2.2 <u>External signal source</u>. The equipment shall be capable of operation with an external signal source, at all frequencies from 50 Hz to 20KHz.

3.11.2.3 <u>Storage factor (Q)</u>. The equipment shall have the capability to measure the storage factor (Q), within the range of 0.05 to 1000, with an accuracy of +/- 5 percent.

3.11.3 <u>Resistance measurement</u>. The resistance measurement capabilities shall extend from 10 ohms to 50 megohms in not more than 8 ranges, with an accuracy of $\pm$ 0.2 percent.

3.11.4 <u>DC bridge excitation</u>.

3.11.4.1 <u>High voltage</u>. High voltage bridge excitation current shall not be less than 10 milliamperes.

3.11.4.2 <u>Low voltage</u>. Low voltage bridge excitation current shall not be less than 250 milliamperes.

4. QUALITY ASSURANCE PROVISIONS

4.1 <u>Responsibility for inspection</u>. Unless otherwise specified in the contract, the contractor is responsible for the performance of all inspection requirements as specified herein. Except as otherwise specified in the contract, the contractor may use his own or any other facilities suitable for the performance of the inspection requirements specified herein, unless disapproved by the Government. The Government reserves the right to perform any of the inspections set forth in the specification where such inspections are deemed necessary to assure supplies and services conform to prescribed requirements.

4.2 <u>Classification of inspections</u>. The inspections required herein are classified as in a and b.

    a. First article inspection (see 4.3).
    b. Quality conformance inspection (see 4.3).

4.3 <u>First article and quality conformance inspection</u>. Unless otherwise specified herein, the first article and quality conformance inspections shall be in accordance with MIL-T-28800.

4.3.1 <u>Test plan</u>. Unless otherwise stated in the contract, the contractor shall prepare a test plan in accordance with paragraphs 4.3(a) and 4.4(a) of MIL-T-28800, for use with both classes of inspection. The test plan shall include as a minimum the tests listed in TABLES I, II, & III, the tests noted in section 4.4, and a description of the satisfactory operation check and satisfactory operation test, as defined in para 4.5 of MIL-T-28800. Unless otherwise required, the tests and inspection to be performed shall be identical for both classes of inspection.

4.3.2 <u>Inspection sampling plan</u>. The inspection sampling to be performed during both classes of inspection shall be as specified in MIL-T-28800, except as indicated below:

4.3.2.1 <u>First production lot</u>. Unless otherwise specified herein or in the contract, the Group C, D, E & F quality conformance inspections shall be performed on the first production lot only.

4.3.2.2 <u>Group D and F tests</u>. The Group D EMI tests and Group F Reliability test shall be performed during quality conformance inspection only if <u>not</u> performed during first article testing.

4.4 <u>Quality assurance tests</u>. Unless otherwise specified herein, the following tests shall be performed in accordance with MIL-T-28800.

4.4.1 <u>Electromagnetic interference</u>. One equipment shall be subjected to an EMI test for compliance with the requirements of 3.9.1. All test set-ups and procedures shall comply with the measurement techniques of Notice 3, MIL-STD-462. All emission and susceptability tests shall be performed with the supplied

TABLE I

Examination and Test Groups

DESCRIPTION	RQMT.	TEST METHOD

Group A
  Preoperational Inspection
  Leakage Current
  Level A Performance
	4.5.3.1*
	3.2.1.3.1*
	TABLE II
		4.3.1

Group B Level B Performance	TABLE III	4.3.1

Group C
  Electrical Power
  Environmental Requirements
  Humidity
  Vibration
	3.5
	3.9
	3.9.2
	3.9.3
		4.3.1

Group D Electromagnetic Interference	3.9.1	4.4.1

Group E
  Dimensions
  Weight
  Front Panel Marking
	3.6.1
	3.6.2
	3.8
		4.3.1
		4.3.1

Group F Reliability	3.10	4.3.1, 4.4.4

*Paragraphs of MIL-T-28800

TABLE II

Level A Performance Tests 1/

DESCRIPTION	REQUIREMENTS
Capacitance Range	3.11.1
Inductance	3.11.2
Resistance Measurement	3.11.3

1/ Level A testing is abbreviated testing (see 6.4.1)

TABLE III

Level B Performance Tests

DESCRIPTION	REQUIREMENTS
DC Test Voltage	3.11.1.1
Dissipation Factor	3.11.1.2
Internal Signal Source	3.11.2.1
External Signal Source	3.11.2.2
Storage Factor (Q)	3.11.2.3

output cable connected to, and extended parallel with, the front of the equipment. The output cable shall be terminated to simulate a normal loading configuration.

4.4.2 Humidity. The equipment shall be subjected to the humidity test specified in para 4.5.5.1.1.3 of MIL-T-28800.

4.4.3 Vibration. The equipment shall be subjected to the vibration tests specified in MIL-T-28800, for Class 5 equipment, except that the equipment shall not be operated during the test. The satisfactory operation test shall be performed prior to and following vibration testing.

4.4.4 Reliability. The reliability tests shall be performed as follows:

4.4.4.1 Burn-in. Each deliverable equipment shall be subjected to a minimum 96 hours on-time burn-in period, prior to Group A testing. During the last 24 hours of burn-in, the equipment must operate failure free. Up until this time, equipment will be allowed to accumulate failures. Each equipment which fails during the final 24 hour period shall be repaired and returned to test until it successfully survives a 24 hour period without failure. Failures which occur during the burn-in test shall be noted and r-ported, but shall not count toward the establishment of equipment MTBF. Prior to burn-in, the satisfactory operation test of 4.3.1 shall be conducted. Daily satisfactory operation checks

(4.3.1) shall be conducted. For the last 24 hour failure free period, a complete satisfactory operation test shall be conducted prior to and after the period.

4.4.4.2 Reliability sampling plan.

4.4.4.2.1 First Article. The Group F reliability demonstration shall be conducted on at least five (5) samples. The use of ten (10) samples is encouraged, but not more than ten (10) samples shall be used.

4.4.4.2.2 Quality conformance inspection. The Group F reliability tests shall be performed on the first production lot, only if not performed during First Article testing. From those units of the first lot that have passed the required Group A and B tests, a random sample of ten (10) units shall be selected, for Group F testing.

4.4.4.2.3 Reliability test plan. The reliability test shall be conducted in accordance with the following test plan:

θ_0 = 2,500 hrs

Number of Failures	Accept** (Equal or More)	Total Test Time* Reject*** (Equal or Less)
0	2000	N/A
1	3750	558
2	5375	2063
3	6875	3850
4	8375	5725
5	9875	7725
6	11,375	11,375****

*   Total test ime is total unit hours of "equipment ON" time (in hours).
**   Accept if test time is greater than or equal to that listed.
***   Reject if test time is less than or equal to that listed.
**** Reject if test time is strictly less than 11,375 with 6 failures.

4.4.4.3 <u>Test length</u>. Testing shall continue until the total unit hours together with the total count of relevant equipment failures permit either an accept or reject decision in accordance with the specified test plan. Only equipment "ON" time may be used in MTBF or longevity determinations. Testing shall be monitored in such a manner that the times to failure may be estimated with reasonable accuracy. No single equipment "ON" time shall be less than one-half the average operating time of all equipments "ON" test.

4.4.4.4 <u>Test conditions</u>.

4.4.4.4.1 <u>Operating mode (duty cycle)</u>. Prior to the burn-in, a satisfactory operation test shall be conducted. Daily satisfactory operation checks shall be conducted. For the last 24 hour failure free period, a complete satisfactory operation test shall be conducted prior to and after the period.

4.4.4.4.2 <u>Test level</u>. While under test, the equipment shall be subjected to the conditions specified by Test Level B, MIL-STD-781.

4.4.4.4.3 <u>Input voltage cycling</u>. When so directed by the procuring activity, voltage cycling shall be accomplished as follows: The input voltage shall be maintained at one hundred ten percent (110%) nominal voltage for one-third of the equipment "ON" cycle, at the nominal value for the second one-third of the equipment "ON" cycle, and at ninety percent (90%) for the final one-third of the equipment "ON" cycle. This cycling procedure is to be repeated continuously throughout the reliability test.

4.4.4.5 <u>Reliability examination and test method</u>. The following inspections shall be used to verify equipment operation during reliability testing. The satisfactory operation test of 4.3.1 shall be used to confirm proper equipment operation prior to and following reliability testing. The satisfactory operation test shall also be performed weekly to verify equipment operation. The satisfactory operation check of 4.3.1 shall be used to monitor proper operation of the equipment daily.

4.4.4.6 <u>Reliability failure actions, Group F</u>. In the event of failure(s) during reliability testing, failure action shall be taken as required in MIL-STD-781, para 5.5.

4.4.4.7 <u>Corrective action, Group F</u>. In the event the reliability test reaches a reject decision, corrective action shall be taken as required in MIL-STD-781, para 5.7.

4.4.5 <u>Optional quality assurance system</u>. The procuring activity may substitute tests from a Department of Defense (DoD) approved supplier's quality assurance system for any or all tests in TABLE I.

5. PACKAGING

5.1 <u>Preservation, packaging</u>. Packaging for delivery shall be in accordance with MIL-T-28800 and as specified by the procuring activity.

6. NOTES

6.1 <u>Intended use</u>. The equipment is intended for use in maintaining electronic communications equipment by equipment repair units in intermediate and depot maintenance.

6.1.1 <u>Equipment replaced</u>. Equipment procured according to this specification is intended to replace all versions of various older equipments now fielded. These equipments to be replaced are listed below:

AN/URM-90	AN/USM-357	ZM-70/U
ZM-30/U	ZM-71/U	

6.2 <u>Ordering data</u>. Procurement documents should specify the following:

    a. Title, number and date of this specification and any amendment thereto.

    b. Packaging requirements (see Section 5).

    c. When rough handling and functional tests are required.

    d. Place of final inspection.

    e. Technical literature required.

    f. Quantity of tools and running spare parts required.

    g. Marking for shipment and shipping containers.

    h. Test plans and test reports.

    i. Classification of inspection and number of samples required.

    j. Rack mounting requirements.

    k. Alternate power requirements.

    l. Maintainablility rationale.

    m. Reliability rationale.

6.3 <u>Contract data requirements</u>.

    a. Nomenclature assignment.

    b. National stock number.

    c. Equipment sample test plans.

    d. Pretest performance records.

    e. First production test data.

    f. Identification plate drawing.

    g. Maintenance Allocation chart.

6.4 Definitions.

6.4.1 Level A performance tests. Level A testing (TABLE II) is a reduced amount of testing which is performed on each equipment produced. Its purpose is to insure that all functions and modes of operation of the equipment are evaluated without extensively checking each parameter as required in the Level B test (TABLE III). The approved equipment test procedure shall specify the actual amount of testing to be performed.

6.4.2 Reliability rationale. Reliability rationale submitted with the bid should provide clear and concise rationale showing how the reliability of the equipment complies with minimum requirements of the solicitation. This may include data from previous reliability tests, reliability predictions and other data available to the offeror. In the absence of such data on the equipment being offered, such data on similar equipment of equal or greater complexity produced by the offeror may be submitted. However, such data must be clearly identified as comparative data, and accomplished with specifications and technical literature on the similar equipment. The acceptance of the submitted rationale does not relieve the successful bidder of performing and successfully completing the production reliability testing.

6.4.3 Maintainability rationale. Maintainability rationale submitted with the bid should provide clear and concise rationale showing how the maintainability of the equipment complies with requirements of the solicitation. This may include data from previous maintainability tests, records of repair and calibration data, and other data on the equipment available to the offeror. Data on selection of module size and other design features relevant to maintainability may also be submitted.

Glossary

Absolute units: Units without prefixes.

Accuracy: A term that describes how close to the true value an instrument can indicate.

Alternate sweep: A method of displaying multiple waveforms on an oscilloscope in which one complete sweep is made of one signal, then of the next signal, and so on, until each signal has been displayed. Then, the process is repeated.

American Standard Code for Information Interchange (ASCII): A seven-bit binary code used to represent 128 letters, numbers, punctuation symbols, and other similar characters.

Amplifier: A circuit in which a small voltage or current controls a large voltage or current.

Analog scale: An instrument scale in which the pointer motion is proportional to the magnitude of the parameter being measured.

Aquadag: A conductive coating within the CRT envelope of an oscilloscope.

Astable multivibrator: A circuit of two mutually exclusive transistors that turn on and off, producing a square wave.

Attenuator: A passive circuit, typically consisting of resistors, that is used as a voltage divider.

Automatic test equipment (ATE): A rapidly growing segment of electronics testing in which multiple measurements are automatically conducted and analyzed by microprocessor-based instrumentation.

Autoranging: A feature in multimeters that allows the range to change automatically. It is controlled by amplifier feedback.

Ayrton shunt: A multirange ammeter circuit in which a shunt is always connected across the meter movement.

Balanced bridge: The condition in an R, L, or C bridge in which the variable controls have been adjusted to the appropriate ratio of the parameter being measured.

Bandwidth: The frequency spectrum between the upper and lower cutoff frequencies.

Bed of nails: A matrix of pins used to connect, through a mask, to a circuit board being tested in an ATE situation.

Block diagram: A symbolic representation of the relationships or flow of information between functional sections of a system.

Bridge: A circuit arrangement with four matched components, used to measure R, L, and C. It has high-quality resistances and/or reactances and measures unknown magnitudes through a process of comparison.

Calibration: The condition of an instrument when its output or response is as close as possible to the specified value. Also, the process of achieving the calibrated condition.

Cathode ray tube (CRT): An electron beam device in a sealed glass envelope, commonly used as the primary display of an oscilloscope.

CGPM: International Committee of Weights and Measures, which considers and recommends improvements in SI units.

Chopped sweep: A method of displaying multiple waveforms on an oscilloscope in which a small portion of one waveform is displayed, then part of the next waveform, and so on, until part of each waveform has been displayed. Then, another part of the first waveform is shown, and so on.

Collimator: An element within a storage CRT used to form a cloud of flood electrons.

Comparator: A digital meter circuit used to recognize ground and input signal levels.

Counter: An electronic instrument used to measure the frequency or period of a signal or to count events as a function of time.

Crest factor: The peak voltage divided by the rms voltage of a given waveform. Mathematically, it describes the variability in one cycle of a signal.

Decade: A device providing resistance, inductance, or capacitance values in steps of 1, 10, 100, and so on.

Decibel: A unit used to express the ratio of two parameters. One decibel is 20 times the base 10 logarithm of the voltage gain or 10 times the base 10 logarithm of the power gain of an amplifier.

Decimal-counting unit: A circuit used in electronic counters to count from 0 to 9. A six-place counter requires six of these circuits.

Deflection factor: The smallest voltage measurable with an oscilloscope.

Deflection sensitivity: The ratio of deflection plate voltage to the beam deflection it produces on the oscilloscope's screen.

Dependent variable: A parameter controlled by another in a relationship. It is plotted on the vertical axis on a graph.

Derived units: Compound units derived from two or more fundamental units. The coulomb is a derived unit; it is derived from the ampere and the second.

Descriptive statistics: Mathematical procedures used to describe and summarize data and produce graphs, charts, and tables. They include the mean, median, and mode.

Device under test (DUT): The component or board being analyzed in an ATE situation.

Dual-beam CRT: A cathode ray tube with two separate and independent sets of electron guns and deflection plates.

Dual-slope ADC: A digital multimeter circuit that measures voltage by integrating the unknown input signal and the known reference signal.

Dual-trace oscilloscope: An oscilloscope that can display two signals at one time by using a chopped or alternate sweep.

Duty cycle: The pulse width divided by the pulse period, expressed as a percentage; an indicator of the on-time of a pulse.

Electrodynamometer: A low-frequency transfer instrument that uses electromagnets.

Error: The difference between true and indicated or assumed values.

Fall time: The time it takes a pulse to fall from 90% to 10% of its maximum value.

Filter: A circuit designed to improve a signal by passing its desired components and removing its undesired components. Typically, it is designed around resistors, inductors, and capacitors.

Flood gun: An element in a storage CRT used to emit flood electrons in order to illuminate a stored image.

FPS: The foot-pound-second system of units.

Frequency response: The effects of frequency variation on the gain, the sensitivity, or other characteristics of a circuit.

Frequency standard: An electronic circuit that produces a high-quality, stable, and precisely known single frequency. It is used as a reference in calibration.

Full-scale deflection (FSD) meter: A meter for which a specified amount of current through or voltage across a meter movement will cause the pointer to move to the maximum limit of the scale.

Function generator: A test instrument that produces sine, square, triangle, and other periodic signals.

Fundamental units: Seven units that serve as the foundation of SI. One fundamental unit is the ampere.

Gain: The ratio of the output voltage or power to the input voltage or power of a circuit or system.

Gate: A circuit of diodes, transistors, or ICs that produces one of two possible levels, according to the input signals and the circuit's logic or rules; the input or controlling element of a field-effect transistor (FET).

General-purpose interface bus (GPIB): A two-directional bus used for communication between signal sources, controllers, and measuring instruments in an ATE situation.

Generator: A test oscillator with modulation capabilities.

Graph: A pictorial display showing the relationship between two or more variables.

Graticule: The lined grid mounted on the face of a CRT to aid in the measurement of deflection.

Hay bridge: An AC bridge circuit used to measure high-Q inductance.

Hysteresis window: The range of insensitivity between limits set for triggering on or off in instruments such as counters. It is used to advantage to ignore small slope reversals.

Impedance bridge: A measurement instrument designed around bridge circuitry and used in the measurement of R, L, and C.

Independent variable: A parameter acting alone or controlling one or more other variables in a relationship. It is plotted on the horizontal axis of a graph.

Inferential statistics: Mathematical procedures used to predict. They include probability based on sampling, mean, and standard deviation.

Instrument: A device used to quantify a parameter.

Instrument error: An error in a measurement caused by the instrument used. For example, an uncalibrated instrument will cause instrument error.

International standard: the highest level of a standard; description based on international agreement and the highest-order physical standard that represents it.

Interval scale: A scale with a zero reference point in which numbers indicate rank order and distance between positions of the scale.

Lab report: A comprehensive document that describes the intentions, procedures, results, and conclusions of a lab investigation.

Limiting error: The result of combining the worst case of all individual errors.

Lissajous pattern: An oscilloscope trace produced by combining two sinusoidal signals, one through the horizontal channel and the other through the vertical channel.

Load effect: A power supply specification that describes the degree to which changes in the supply load can affect the output voltage.

LVDT: A linear variable differential transformer, a transducer that produces a voltage as a function of core position.

Manipulator: Part of a robot, often capable of moving a tool or gripper to various positions through linear and/or angular motion.

Maxwell bridge: An AC bridge circuit used to measure low-Q inductance.

Mean: The average of a group of magnitudes.

Measurement: The process of quantifying a parameter.

Measures of central tendency: Mathematical descriptions of the grouping of individual measures.

Median: The middle magnitude in a group of magnitudes placed in rank order.

Meter loading: A measurement condition in which an instrument affects the operation of the circuit being measured. Typically, it is a result of a low-sensitivity voltmeter drawing its deflection power from, and reducing the voltage at, the point being measured.

Midrange: The magnitude in a group of magnitudes located midway between the largest and the smallest.

MKS: The meter-kilogram-second system of units, which is the foundation of SI.

Mode: The most frequently occurring magnitude in a group.

Modulation: The process of combining or mixing information and a carrier, such as combining AF onto an RF carrier.

Monostable multivibrator: A transistor-switching circuit that produces one pulse for each input command.

Multiplier: A series resistance used to extend the range of a voltmeter.

NBS: National Bureau of Standards, which is responsible for maintenance of standards in the United States.

Network analyzer: A measurement device capable of providing a programmed sequence of input signals to another device as well as measuring and analyzing that device's response.

Nominal scale: A scale in which numbers are used only for identification or classification, as in a telephone number.

Normal distribution curve: A graph showing the number of occurrences of an event (vertical axis) as a function of the magnitude of the event (horizontal axis) in normal populations. It is also called a bell curve because of its shape.

Null: A condition in measurement when two magnitudes being compared are equal.

Null balance: The status in a bridge measurement process when the ratio of the internal known value matches the external unknown value.

Null detector: A device, such as a meter or an oscilloscope, used to indicate the null condition.

Observer error: The error in a measurement caused by the person conducting the measurement. Examples are misreading a dial and selecting an inappropriate instrument.

Offset voltage: A positive or negative DC voltage internally added to the output signal of a function generator.

Operational amplifier: A multipurpose amplifier circuit with high-impedance differential inputs and a low-impedance output. Typically found on an integrated circuit, it can be used as a high- or low-gain, inverting or noninverting amplifier, oscillator, regulator, or active filter.

Ordinal scale: A scale in which numbers indicate rank or order.

Oscillator: A circuit or system that produces a sinusoidal output.

Oscillator-divider: A circuit used in counters and digital meters. A crystal-controlled oscillator produces a specific and stable frequency, which is then reduced, in multiples of a tenth, to the desired frequency.

Parallax error: An error in the interpretation or the reading of a scale that occurs when the viewer is not aligned correctly in the intended viewing position.

Parameter: A circuit dimension such as voltage or resistance.

Parameter error: An error in a measurement caused by abnormal phenomena in the parameter, such as transient voltages.

Periodic and random deviation (PARD): A power supply specification that describes the quality or smoothness of the output voltage.

Persistence: The duration of CRT image retention.

Phase-locked loop (PLL): A closed-loop circuit consisting of a voltage-controlled oscillator, phase detector, low-pass filter, and amplifier. It utilizes this circuitry and an external reference frequency to provide a high-quality, stable output signal.

Phase relationship: The number of degrees between the points where two signals of the same frequency cross zero magnitude at a positive slope.

Phase-shift oscillator: A device that produces a sinusoidal signal by taking the output of an active circuit, such as an amplifier, shifting that output by 180 degrees, and feeding it back to the amplifier input, causing self-sustaining oscillation.

Physical standard: A physical representation of the defined standard for a parameter.

Potentiometer: A precision instrument used for measuring small DC voltages by comparison.

Precision: The degree of resolution or number of significant figures an instrument will give.

Prefix: A word used to represent a unit multiplier, such as *kilo,* which is used to represent "times 1000."

Primary standard: The fundamental definition or physical representation of a unit kept by an organization or a country.

Probability: The mathematical description of the likelihood of a magnitude or event occurring. It is based on the prior observation of earlier occurrences.

Pulse width: The time elapsed while a pulse is between 50% of maximum on its rise and 50% of maximum on its fall.

Ramp ADC: A circuit commonly used in digital voltmeters in which the unknown voltage is compared with a ramp produced by a charging *RC* circuit.

Random error: An error that is unpredictable and inconsistent.

Ratio resistor: An internal resistor used for comparison with an unknown in a measurement circuit, such as a Wheatstone bridge circuit.

Ratio scale: A scale in which numbers indicate order and indicate distance between each other. It has a zero reference point that indicates complete absence of the quantity.

Recorder: An electromechanical instrument that produces a permanent visual record of one or more parameters as a function of another, such as time.

Rectifier: A circuit that changes AC voltage to pulsating DC voltage.

Regulator: A circuit used in power supplies to stabilize the output voltage. It reduces the effects of changes in input voltage and output demand.

Reliability: The repeatability or degree of consistency of an event, magnitude, or instrument reading.

Retrace blanking: An oscilloscope circuit that turns off the CRT beam as it returns from right to left to begin another sweep.

Ripple: A residual AC component in the output of a DC power supply. It is undesirable, and filters are used to remove it.

Ripple factor: The ratio of AC rms voltage divided by the DC output voltage of a power supply.

Ripple frequency: The frequency of the AC component in the output of a DC power supply.

Rise time: The time it takes a pulse to rise from 10% to 90% of its maximum value.

Robot: A programmable electromechanical device used to move objects through routine and repeated patterns.

Sawtooth frequency: The frequency of the internal horizontal circuit in an oscilloscope, a circuit that is used to move the beam from left to right.

Schematic: A symbolic representation showing the interconnection of specific components in a circuit or system.

Schering bridge: An AC bridge used to measure low-*D* capacitances.

Schmitt trigger: A circuit used to produce square waves from varying signals such as sine waves.

Secondary standard: A close duplication of a primary standard, kept by manufacturers and used for comparison with working standards.

Sensitivity (meter): The ratio of movement resistance to full-scale deflection voltage, expressed in ohms per volt.

Servo-balancing system: A circuit in which an amplifier-motor system adjusts a controlled element to follow a reference element.

Shunt: A parallel resistance used to extend the range of an ammeter.

Shunt box: A resistive divider and shunt circuit used to convert a current to a voltage measurable by a potentiometer.

SI: International System of Units.

Signal sampling: A process used in oscilloscopes to measure signals with frequencies beyond the normal range.

Significant figures: The degree of resolution or certainty to which a magnitude is known.

Source effects: A power supply specification that describes the degree to which changes in the input voltage can affect the output voltage.

Spectrum analyzer: A measurement device that provides a varying-frequency input signal to a circuit and displays its output as a function of frequency.

Standard: A universally accepted description or definition of a unit's magnitude.

Standard cell: A glass-enclosed voltage source used as a reference for calibration.

Standard deviation: The root-mean-square value of variation of a group of magnitudes around their mean.

Storage oscilloscope: An oscilloscope with the capability of displaying an image for an extended time or for storing an image in the CRT for later display.

Strain gage: A resistive strip that produces a resistance change as a function of extension (strain).

Strip chart recorder: A recorder that utilizes a roll of graph paper to produce a permanent record of the relationship between one or more parameters as a function of time.

Successive-approximation ADC: A circuit used in a digital voltmeter. An incoming signal is sampled and compared with a succession of progressively smaller reference signals in order to produce a digital description.

Sweep circuit: An oscilloscope circuit that produces the signal for moving the beam from left to right in a predetermined period of time.

Sweep oscillator: An instrument used in component and system evaluation. It produces sinusoidal signals with adjustable amplitude that can be manually or automatically varied over a selected frequency range.

Systematic error: An error that follows a pattern and is predictable, such as the effects of temperature change.

Table: Columns of data arranged to show the relationships between independent (left column) and dependent (inside columns) variables.

Test: A process of observation or examination.

Thermocouple: A joined pair of dissimilar conductors that produce a predictable voltage proportional to temperature.

Time base: A crystal-controlled circuit in a counter used in conjunction with dividers to produce a specific on-off time for frequency or event counting.

Traceability: A method of documentation used to trace the accuracy of measurements back to those kept by the National Bureau of Standards.

Transducer: An electric or electronic device that converts energy from one form to another.

Transfer instrument: An instrument that can measure the effective value of both DC and AC current directly, regardless of waveform.

Transformer: A passive component with coils and core used to change AC voltage level, to match impedance, and to isolate grounds in circuits.

Trigger: An oscilloscope circuit or signal used to command the beginning of a horizontal sweep.

Unbalanced bridge: A condition of an R, L, or C bridge prior to final adjustment of the reference ratios.

Unit: The name of a parameter's dimension, such as meter for length and second for time.

Variability: The differences in magnitude of individual measurements of a given parameter.

Variable-frequency oscillator: An oscillator in which frequency is controlled, and varied, by variable capacitance, resistance, voltage, or some other parameter.

Voltage divider: A series resistive circuit used to provide reduced magnitudes of the total applied voltage.

Volt box: A resistive divider used to reduce DC voltages to a range measurable by a potentiometer.

Waveform error: An error in the averaging type of AC meters as they respond to nonsinusoidal waveforms.

Wheatstone bridge: A DC bridge circuit used to measure resistance.

Wien bridge oscillator: An oscillator circuit using series and parallel RC branches to control frequency.

Working standard: A standard used in the calibration of instruments.

Write gun: An element that emits and directs the beam to be stored in a storage oscilloscope CRT.

WWV and WWVB: Transmitters of the NBS, providing carriers and tones for use as time, frequency, and voltage standards.

X–Y recorder: A chart recorder that produces a single-sheet graph of one or more parameters as a function of another.

Answers to Selected Problems

Chapter 2

1. 2.4×10^{-4} ampere, 7.5×10^5 ohms, 2×10^{-8} farad, 3.5×10^7 ohms
4. 1500 volts, 0.150 volt, 0.00000472 volt, 470,000 ohms
7. 350 microamperes, 0.750 kilovolt, 0.0022 microfarad

Chapter 3

1. 10 watts, 8.1%
2. 160 ohms, 8.4%
6. mean = 20.0, median = 19.1, mode = 18.6, midrange = 19.0
8. standard deviation = 4.43 volts, 50% = 17.03 to 22.97, 95% = 11.14 to 28.86, 99% = 6.71 to 33.29
10. Points from -3σ to $+3\sigma$ are as follows: 6.71, 11.14, 15.57, 17.03, 20.0, 22.97, 24.43, 28.86, 33.29
12. 14.9 to 25.1 volts ($\pm 1.15\sigma$ for 37.5% of area each side of mean)

Chapter 5

1. 8 kilohms, 498 kilohms, 1.498 megohms
5. 500 ohms, 105.3 ohms, 20.2 ohms
7. 1 kilohm, 1 kilohm, 3 kilohms
9. 5 kilohms, 10 kilohm pot set at 5 kilohms
11. 5 kilohms, 10 kilohm pot set at 5 kilohms
13. 22.5 kilohms, 7.5 kilohms, 2.5 kilohms
15. 1993.6 ohms, 3333.3 ohms, 10 kilohms
17. 9.38%, 4.69%
19. 4.5 watts, 14.8%
21. 33.3%

Chapter 6

1. Change divider to 9 megohms, 0.9 megohm, 90 kilohms, 9 kilohms, and 1 kilohm.
3. 5 megohms, 4 megohms, 500 kilohms, 400 kilohms, 50 kilohms, 50 kilohms
7. 13.2 volts
11. 1.337, 1.366

Chapter 7

1. 275 ohms
5. 3 henrys, 300 ohms, 62.8
8. 0.787 henry, 622 ohms, 7.95
11. 0.015 microfarad, 100 ohms, 0.00943
14. 0.5 microfarad, 100 ohms

Chapter 8

1. 20 volts per centimeter
3. 250 millivolts peak to peak, 88.4 millivolts rms, 1 millivolt peak to peak, 0.354 millivolt rms
4. 200 millivolts DC, 0.8 millivolt DC
7. 80 milliseconds, 12.5 hertz, 1 millisecond, 1 kilohertz
9. 41.8 degrees
11. 45 degrees

Chapter 9

1. 64,974 hertz
5. 19,894.3 hertz
9. 41.6 decibels
11. 17 decibels
13. 50%
17. 0.1 millisecond, 30%

Chapter 11

1. 13.8%
4. 0.25%

7. center-tapped transformer with turns-ratio of 4.5:1
8. DC output rises to 13.3 V DC while ripple decreases to 3.9 V rms.
11. $R_1 = 280$ ohms, $R_2 = 133 + 267$ ohms

Index